W0107582

Induced Seismicity

Edited by
Harsh K. Gupta
Rajender K. Chadha

1995

Birkhäuser Verlag
Basel · Boston · Berlin

Reprint from Pure and Applied Geophysics
(PAGEOPH), Volume 145, No. 1

The Editors:

Dr. Harsh K. Gupta
Director
National Geophysical Research Institute
Uppal Road
Hyderabad 500007
India

Dr. Rajender K. Chadha
Scientist
National Geophysical Research Institute
Uppal Road
Hyderabad 500007
India

A CIP catalogue record for this book is available from the Library of Congress, Washington D.C., USA

Deutsche Bibliothek Cataloging-in-Publication Data

Induced seismicity / ed. by Harsh K. Gupta; Rajender K. Chadha. – Basel ; Boston ; Berlin : Birkhäuser, 1995

ISBN-13: 978-3-7643-5237-0   e-ISBN-13: 978-3-0348-9238-4
DOI: 10.1007/978-3-0348-9238-4

NE: Gupta, Harsh K. [Hrsg.]

© 1995 Birkhäuser Verlag, P.O. Box 133, CH-4010 Basel, Switzerland
Printed on acid-free paper produced from chlorine-free pulp

9 8 7 6 5 4 3 2 1

Contents

PAGEOPH, Vol. 145, No. 1 (1995)

0033–4553/95/010001–02$1.50 + 0.20/0

Introduction

The first known example of reservoir-induced seismicity was documented at Lake Mead reservoir, created by the Hoover Dam on the Colorado river in the United States of America during the late 1930s. During the 1960s several earthquakes exceeding magnitude 6 occurred in the vicinity of artificial water reservoirs at Xinfengjiang, China (1962); Kariba, Zambia-Zimbabwe border (1963); Cremasta, Greece (1966); and Koyna, India (1967). Over the years, the number of cases of induced earthquakes as a consequence of some of man's engineering activity, such as the impounding of deep artificial water reservoirs, underground mining,. high-pressure fluid injection, large-scale surface quarry, oil production, geothermal power generation, waste disposal, etc. has increased and crossed 100 known sites globally. The Koyna reservoir in India continues to be the most significant site of reservoir-induced earthquakes. Such earthquakes began to occur soon after the impoundment of the Shivajisagar Lake in 1962 and continue to occur, the latest being the 12th and 13th of March, 1995 earthquakes exceeding magnitude 4. With the advent of sophisticated instrumentation, borehole measurement, and advanced computational techniques, it appears that we are closer to apprehending the mechanism of induced earthquakes, and thereby improving our understanding of the physics of the earthquakes.

Keeping in view the importance of phenomena of induced earthquakes, a workshop on 'Induced Seismicity' was organized during the 27th General Assembly of the International Association of Seismology and the Physics of the Earth's Interior held at Victoria University, Wellington, New Zealand from 10–21 January, 1994. For this workshop a total of 40 abstracts were received from participants from 16 countries. The present volume includes 16 papers selected from these presentations made at the workshop and a few other papers which, in the view of the editors, deserve inclusion in the volume.

The first 3 papers deal with mining activity induced seismicity, the 4th paper pertains to induced seismicity caused by oil extraction, while the remaining 12 papers focus on various aspects of artificial water reservoir induced earthquakes.

In the first paper Trifu *et al.* examined the source parameters of mining induced earthquakes based on homogeneous and inhomogeneous models and they emphasized the importance of an inhomogeneous model for the assessment of seismic hazard in mining regions. Kremenetskaya and Trjapitsin, in the second paper, report the enhancement of seismicity after removing a large amount of rock mass in the Kola Peninsula. They suggest the possibility of earthquakes being triggered by explosions used for mining purposes. The third paper in this category is by McKavanagh *et al.* who point out the importance of monitoring rock and quarry blasts using regional seismic networks. They have successfully used triaxial seismometers to record seismic events produced by the strata failure and roof falls in coal mines. They believe that initial fall under massive roof conditions appears to be predictable by following the trend of microseismic events and longwall production rate. The fourth paper by Zhao Gen Mo *et al.* deals with the study of earthquakes induced by the water injection at four oil fields in China. The controlled experiment of water injection showed the relationship between the water injection and the consequent earthquake. The stress drop is found to vary from 0.2 to 3.0 bars and the Q factor is estimated to be 75. They also note that in the Renqiu oil field, the earthquakes induced by water injection are characterized by low stress drop.

The remaining 12 papers of this volume address various aspects of reservoir-induced earthquakes. Four papers by Awad and Megume Mizoue, Kusala Rajendran, Piccinelli *et al.* and Rastogi *et al.* concern seismicity monitoring near reservoirs and the study of induced earthquake characteristics vis-à-vis reservoir water levels. The paper by Shen Li-Ying and Chang Bao-Qi explains the induced seismicity at the Xinfengjiang reservoir in China by the interaction of stresses and the pore pressure changes. Based on laboratory studies, Panfilov and Sobolev suggest that under favorable P-T conditions the unstable steam-water convection can trigger seismicity. Deyi Feng *et al.*, based on their study of 15 cases of reservoir-induced phenomenon from different countries, underscore the utility of the Fuzzy Multifactorial Evaluation method to assess the potential strength of induced earthquakes. The next three papers by Chadha, Talwani and Raval deal with the geological and seismotectonic factors which may contribute to the initiation of induced seismicity in a reservoir region. In another paper, Awad and Mizoue correlate the deeper seismic activity in the Aswan region in Egypt with a high-velocity anomaly derived from a tomography study. The concluding paper by Srivastava *et al.* attempts a novel approach to study reservoir associated characteristics using Deterministic Chaos.

For the successful completion of this volume we wish to thank, Renata Dmowska, the Co-Editor in Chief, Pure and Applied Geophysics for inviting us to be the Guest Editors, the authors for contributing high quality papers, and to the conscientious reviewers who played an important role in the review process. We are confident that continuous research in the area of 'Induced Seismicity' will contribute greatly to our quest towards understanding the earthquake generating processes which will help in achieving the ultimate goal of earthquake prediction.

Harsh K. Gupta
R. K. Chadha

PAGEOPH, Vol. 145, No. 1 (1995)

0033–4553/95/010003–25$1.50 + 0.20/0
© 1995 Birkhäuser Verlag, Basel

Source Parameters of Mining-induced Seismic Events: An Evaluation of Homogeneous and Inhomogeneous Faulting Models for Assessing Damage Potential

CEZAR-IOAN TRIFU,[1] THEODORE I. URBANCIC,[1] and R. PAUL YOUNG[2]

Abstract—Source parameter estimates based on the homogeneous and inhomogeneous source models have been examined for an anomalous sequence of seven mine-induced events located between 640 and 825 m depth at Strathcona mine, Ontario, and having magnitudes ranging between m_N 0.8 and 2.7. The derived Brune static stress drops were found to be similar to those observed for natural earthquakes (~30 bars), whereas dynamic stress drops were found to range up to 250–300 bars. Source radii derived from Madariaga's model better fit documented evidence of underground damage. These values of source radii were similar to those observed for the inhomogeneous model. The displacement at the source, based on the observed attenuation relationship, was about 60 mm for three magnitude 2.7 events. This is in agreement with slip values calculated using peak velocities and assuming the asperity as a Brune source within itself (72 mm). By using Madariaga's model for the asperity, the slip was over 3 times larger than observed. Peak velocity and acceleration scaling relations with magnitude were investigated by incorporating available South African data, appropriately reduced to Canadian geophysical conditions. The dynamic stress drop scaled as the square root of the seismic moment, similar to reported results in the literature for crustal earthquakes. This behavior suggests that the size of the asperities responsible for the peak ground motion, with respect to the overall source size, follow distributions that may be similar over a wide range of magnitudes. Measurements of source rupture complexity (ranging from 2 to 4) were found to agree with estimates of overall source to asperity radii, suggesting, together with the observed low rupture velocities ($0.3\,\beta$ to $0.6\,\beta$), that the sources were somewhat complex. Validation of source model appropriateness was achieved by direct comparison of the predicted ground motion level to observed underground damage in Creighton mine, located within the same regional stress and geological regime as Strathcona mine. Close to the source (<100 m), corresponding to relatively higher damage levels, a good agreement was found between the predicted peak particle velocities for the inhomogeneous model and velocities derived based on established geomechanical relationships. The similarity between asperity radii and the regions of the highest observed damage provided additional support for the use of the inhomogeneous source model in the assessment of damage potential.

Key words: Mining-induced seismicity, faulting models, peak ground parameters.

[1] Engineering Seismology Group Canada Inc., Kingston, Canada, K7L 2Z4.
[2] Department of Geology, University of Keele, Staffordshire, UK, ST5 5BG.

Introduction

Traditionally, estimates of seismic source parameters, such as source dimensions and stress release, are obtained from well-known homogeneous source models as proposed by BRUNE (1970; 1971) and MADARIAGA (1976). These models assume a uniform stress release over the entire source area. Brune's model is based on a circular dislocation along which an instantaneous stress drop occurs, whereas Madariaga's model considers the source as an expanding circular crack with finite rupture velocity, therefore allowing for variations in pulse shape with orientation from the source to be taken into account. However, some evidence in the literature suggests that these models may not always be appropriate. For example, estimates of source radius have been found to be too large relative to observed geological evidence (GIBOWICZ, 1984). BOATWRIGHT (1984) attempted to ascertain the effect rupture geometry has on source size by assuming that the overall source is composed of segments, or sub-events, that are separated by barriers or regions without stress release. For events with magnitudes of around 4, he found that the number of sub-events ranged from 1 to 2.5, suggesting that these earthquakes had moderately complex geometries. McGARR (1981) compared ground motion parameters for tremors in a deep gold mine with values inferred from crustal earthquakes, and proposed an inhomogeneous model of the seismic source. According to this model, the stress drop is not uniformly released over the entire source area, but is confined to asperity-like patches. Since different source models predict different values of ground motion, the selection of a specific model is therefore critical when assessing the seismic hazard within an earthquake source region. This is of particular concern when considering the placement and maintenance of underground facilities, such as in a mining environment. It is the goal of this study to evaluate the appropriateness of homogeneous or inhomogeneous source models for assessing damage potential associated with mining-induced earthquakes.

The mines of the Sudbury Basin, Ontario provide a suitable environment for recording substantial tremors that have the potential to affect existing and proposed underground facilities. During June, 1988 an anomalous sequence of seven events having Nuttli magnitudes (NUTTLI, 1973) ranging between 0.8 and 2.7 was recorded over a two-day period in Strathcona mine. The events occurred between 640 and 825 m below surface, as located with a 16-channel uniaxial/triaxial seismic network local to the mine, operated by Queen's University (URBANCIC and YOUNG, 1990). Strong motion recordings were obtained at distances of several hundred meters with a five station triaxial network operated by the Canmet Laboratories (MAKUCH, 1986). The present study examines source parameters based on homogeneous and inhomogeneous source models for the above data set. These include source and asperity radii, slip and slip velocity, static and dynamic stress drops. Source rupture complexity analysis is used to provide additional evidence for the selection of either a homogeneous or inhomogeneous source model.

Observed peak particle velocity and acceleration scaling relations with magnitude, as determined by integrating our data with that available in the literature (MCGARR et al., 1981), are explored in order to determine the appropriateness of different source models. The implications of the observed scaling relations to the strength distribution on the fault and high frequency radiation at the source are also discussed. Finally, we consider the applicability of these models to the evaluation of damage potential associated with mining-induced seismic events.

Instrumentation

Two independent monitoring systems were located at the Strathcona mine during June, 1988. The Queen's system, a 14 sensor array, consisting of 13 uniaxial accelerometers and a CE 507 M301 Vibrometer triaxial accelerometer, was used primarily to record microseismic events (magnitude less than zero). The array sensors were uniformly distributed within a $200 \times 200 \times 200$ m volume at about 750 m depth. The uniaxial sensors had a frequency response of 1 to 10 kHz with a total sensitivity of 950 V/g, whereas the triaxial sensor had a flat response from 50 Hz to 5 kHz and a total sensitivity of 6 V/g. Data acquisition was carried out with 12-bit A/D conversion at a sampling rate of 20 kHz per channel. The main function of the Queen's system was to provide accurate locations based on P-wave arrivals (see Table 1) and additional P-polarity measurements for fault-plane solutions. Arrival times were picked to within 0.15 ms at stations having signal-to-noise ratios greater than 4. Locations were obtained based on at least 10 arrivals, with typical hypocentral distances of less than 100 m, through a combined Simplex-Geiger algorithm (GENDZWILL and PRUGGER, 1985; THURBER, 1985) to an accuracy of about 5 m.

Table 1

The events analyzed in this study. Magnitudes were obtained from the Geological Survey of Canada, with the exception of event number 4, which was derived from measurements of seismic moment with the Canmet system.

No.	Date (DDMMYY)	UTC (HH:MM:SS)	Depth (m)	m_N
1	190688	09:53:21	678	2.7
2	190688	10:20:04	748	2.4
3	190688	10:43:13	643	1.8
4	200688	20:44:04	820	0.8
5	210688	00:41:35	814	2.0
6	210688	01:44:52	814	2.7
7	210688	05:08:16	825	2.7

The second system, established by the Canmet Laboratories, consisted of a 5 triaxial station network (S100 Teledyne-Geotech accelerometers) used to record strong motions. These sensors had a flat frequency response (within 3 dB) between 0.1 Hz and 1 kHz, and a sensitivity of 5 V/g. Overall, the system input was limited to 2 g. The actual system response was band-limited from 1 to 150 Hz with a recursive Butterworth bandpass filter (-96 dB/octave drop-off). Digitization frequency of the analog signals was set to 1024 Hz. Three sensors were installed at the surface. These were placed in 6 m deep boreholes in order to minimize cultural noise on the recorded signals. Source-sensor separations ranged from 879 to 1317 m. The remaining two sensors were located within the mine, at 724 and 838 m depth. These were also installed in boreholes to reduce the influence of openings and mine noise. The source-sensor separations for these sensors ranged from 87 to 425 m.

Source Parameter Determination Using a Homogeneous Source Model

The homogeneous source model states that an earthquake is generated by the sudden release of shear stress over an entire area of the fault surface. In this study, calculations of source parameters were made using far-field amplitude spectra for both the P and S waves based on the models of BRUNE (1970; 1971) and MADARIAGA (1976). The data analysis was carried out on Canmet signals rotated from the recording coordinate system to the local ray coordinate system based on linear P-wave polarizations (MATSUMURA, 1981). All rotated signals had linearities greater than 90%; this provided for the accurate identification of S-wave phases. The spectral analysis was performed with an interactive graphics package that allowed for the categorization of the signal type (P, S, or noise), up to two spectra to be simultaneously displayed, and with user specification of the FFT window length. A 10-point cosine taper was applied to the time signals to remove possible edge effects. Two spectral windows were defined for each processed waveform; one over the P or S wave and a secondary window over the background noise preceding the P-wave signal. Displacement and acceleration spectra were corrected for instrument response, scattering, and anelastic attenuation. Corrections for instrument response included limiting the data to the instrument recording frequency bandwidth and accounting for the system amplification response. To correct for scattering and anelastic attenuation, we considered the exponential term

$$F_a = \exp(\omega t^*) = \exp(2\pi f R/cQ) \tag{1}$$

where t^* is the attenuation operator, f is the frequency, R is the source-sensor separation, c is either the P-wave velocity $\alpha = 6095$ m/s or the S-wave velocity $\beta = 3800$ m/s, and Q is the average quality factor. An assumption that $Q_\alpha = (9/4)Q_\beta$ (BURDICK, 1978) was accepted and a value of $Q_\beta = 100$ was used as being typical for underground hard rock environments (e.g., GIBOWICZ *et al.*, 1990). A displace-

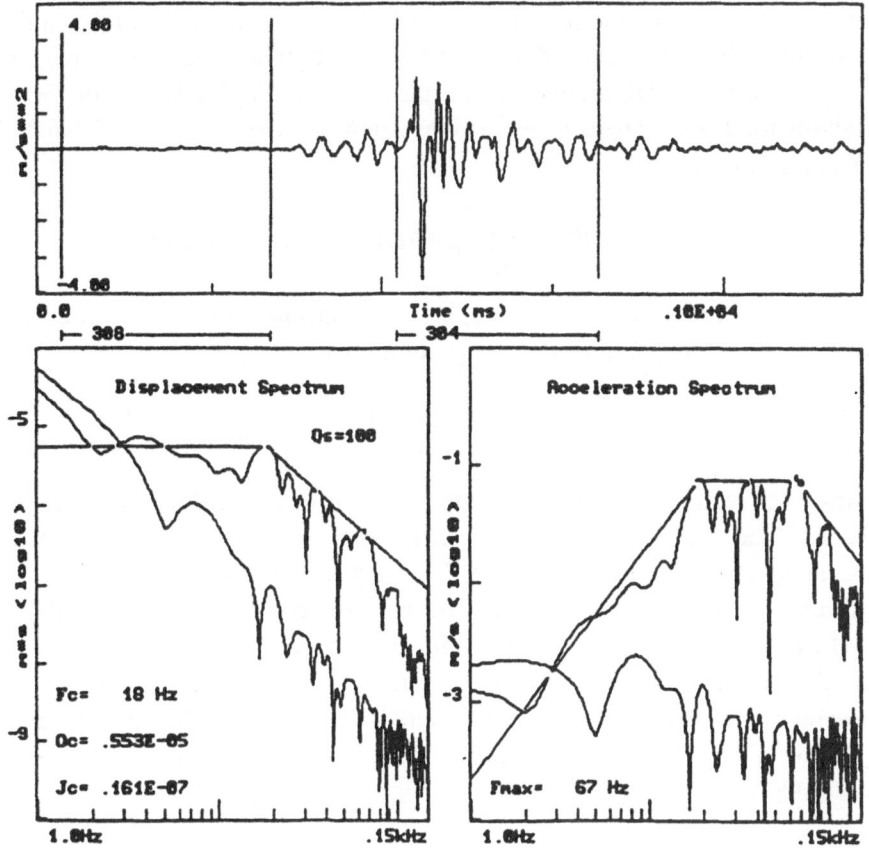

Figure 1

Example of rotated S-wave attenuation corrected displacement and acceleration spectra, and the uncorrected noise spectra for event 2 at a distance of 1290 m. Also shown are the corner frequency Fc (Hz), spectral level Oc (m · s), and energy flux Jc (m²/s). The frequency range corresponds to the filtered bandwidth (1 to 150 Hz), and the FFT noise and S-wave windows are as indicated (308 and 304 sample points, respectively).

ment spectral fall-off of ω^{-2} to ω^{-3} was observed for uncorrected spectra, whereas corrected spectra were well described by a ω^{-2} fall-off (Figure 1).

Two independent parameters were calculated directly from the displacement spectra, the low frequency level (Ω_{oc}) and the energy flux (J_c). Automatic estimates of Ω_{oc} were obtained as proposed by ANDREWS (1986)

$$\Omega_{oc}^2 = 4S_{D2}^{3/2} S_{V2}^{-1/2} \tag{2}$$

with

$$S_{D2} = 2\int_0^\infty D^2(f)\, df, \quad S_{V2} = 2\int_0^\infty V^2(f)\, df \tag{3}$$

where D^2 and V^2 denote the squared displacement and velocity spectral amplitudes, respectively. The calculation of energy flux, as obtained according to Snoke's algorithm (SNOKE, 1987), assumes a constant spectral amplitude Ω_{oc} for $f < f_1$ and f^{-2} fall-off for $f > f_2$, where f_1 and f_2 correspond to the instrumental bandwidth. The relation used was

$$J_c = \tfrac{2}{3}|\Omega_{0c}\omega_1|^2 f_1 + 2\int_{f_1}^{f_2} |\omega U(\omega)|^2 \, df + 2|\omega_2 U(\omega_2)|^2 f_2 \tag{4}$$

where $U(\omega)$ is the displacement amplitude spectrum. Based on these spectral parameters, the corner frequency f_c was computed as

$$f_c = \left(\frac{J_c}{2\pi^3 \Omega_{oc}^2}\right)^{1/3}. \tag{5}$$

Spectral level corner frequency estimates were further visually inspected for discrepencies. The observed corner frequencies were within the instrumental bandwidth, ranging from about 10 to 40 Hz. The frequency above which the acceleration spectra of S waves show a sharp decrease with increasing frequency is referred to $f_{\max}$ (HANKS, 1982). Values were determined by approximating the intersection of the acceleration plateau with the high frequency slope.

Several source parameter estimates were obtained from the spectral analysis. The seismic energy E_c was calculated directly from the energy flux J_c as outlined by SNOKE (1987)

$$E_c = \frac{4\pi\rho c R^2 J_c}{F_c^2} \langle F_c^2 \rangle \tag{6}$$

where ρ is the density of the source material (2700 kg/m³), F_c is the radiation pattern coefficient, and $\langle \rangle$ represent the average values. We have assumed that the loss in energy from scattering and anelastic attenuation has been taken into account, and that the individual radiation pattern coefficients are similar to average values. In our case, the average coefficients obtained from individual fault-plane solutions were 0.33 for P waves and 0.63 for S waves. These are similar to those proposed by MCGARR (1991), with 0.39 and 0.57 for P and S waves, respectively. In order to allow for a direct comparison of results, the later values were used throughout this study.

The P- and S-wave seismic energies for the investigated events are as provided in Table 2. It is worth noting that the relative S-wave to P-wave energy ratios are smaller (<10) than typically assumed for a pure shear failure (BRUNE, 1970; BOATWRIGHT and FLETCHER, 1984). This suggests that a non-shear volumetric component likely exists as part of the failure mechanisms. However, the fit of the double-couple model to observed polarity measurements implies that shear remains the dominant mode of failure (Figure 2). For the purposes of this study, only the

Table 2

Source parameters as deduced from spectral analysis using Brune's and Madariaga's models.

No.	M_0 (dyne · cm)	r_0 (m)	r_0^M (m)	$\Delta\sigma$ (bar)	$\Delta\sigma_{\mathrm{rms}}$ (bar)	σ_{app} (bar)	E_α (J)	E_β (J)
1	8.1 (10^{19})	84	47	61	21	27	4.8 (10^7)	5.7 (10^8)
2	3.1 (10^{19})	78	44	29	14	27	5.4 (10^7)	1.3 (10^8)
3	3.4 (10^{18})	50	28	10	5	10	3.8 (10^6)	9.1 (10^6)
4	9.8 (10^{17})	39	22	7	2	7	8.3 (10^5)	1.7 (10^6)
5	9.1 (10^{18})	55	31	22	10	14	1.5 (10^7)	3.3 (10^7)
6	7.8 (10^{19})	114	64	23	12	14	1.5 (10^8)	2.9 (10^8)
7	8.8 (10^{19})	93	52	48	22	16	8.0 (10^7)	3.6 (10^8)

shear component of failure was considered in further source parameter determinations.

The seismic moment M_0 was evaluated by

$$M_0 = \frac{4\pi\rho\beta^3 R |\Omega_0|}{F_\beta} \tag{7}$$

where M_0 is defined for the vector sum of the S-wave components, $M_0 = (M_{0,SV}^2 + M_{0,SH}^2)^{1/2}$. Estimates of source radius (r_0, r_0^M) were obtained by assuming

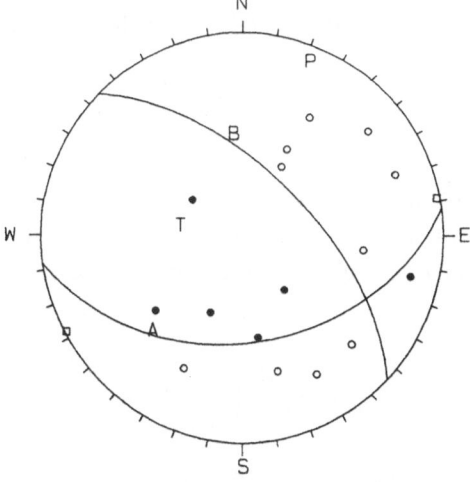

Figure 2

Typical fault-plane solution (lower hemisphere) as obtained for event 2 where dilatational first motions are represented as open circles; compressional first motions are closed circles; A and B correspond to the nodal planes; and P, T are the pressure and tension axes, respectively.

Brune's and Madariaga's models, respectively. For a circular fault, the source radius is inversely related to the corner frequency by the constant K_β,

$$(r_0; \, r_0^M) = \frac{K_\beta^{(B;M)}\beta}{2\pi f_\beta}.$$ (8)

In assuming Brune's model, the calculated source radii are approximately twice those based on Madariaga's model. These differences result from the K_β coefficient value relating corner frequency to source radius in the two models, $K_\beta^B = 2.34$ compared to average $K_\beta^M = 1.32$ for an assumed rupture velocity of $0.9\,\beta$. Under the assumption of a homogeneous model, K_β^M lead to source radii that better fit observed damage underground than values based on K_β^B (to be subsequently discussed in the section on damage potential).

In this study, three estimates of the average stress release on the faulting surface were made, namely the static stress drop, the apparent stress, and the rms stress drop. The static stress drop $\Delta\sigma$, which represents the average stress release over the entire source, was calculated as

$$\Delta\sigma = \frac{7}{16}\frac{M_0}{r_0^3}.$$ (9)

The values of $\Delta\sigma$ listed in Table 2 use source radii based on Brune's model; in assuming Madariaga's model the $\Delta\sigma$ values would be $[K_\beta^B/K_\beta^M]^3 \approx 5.57$ times larger. Our static stress drop values ranged from 10 to 50 bars, consistent with reported estimates for worldwide earthquakes. An independent estimate of the average shear stress drop during a seismic event is obtained by the apparent stress σ_{app}

$$\sigma_{app} = \frac{\mu E_\beta}{M_0}$$ (10)

where $\mu = \rho\beta^2$. Finally, the rms stress drop σ_{rms}, which is defined as the stress release from all subevents in the faulting process, is calculated from measurements of the rms acceleration a_{rms} following the approach established by HANKS and MCGUIRE (1981)

$$\sigma_{rms} = \frac{2.7\rho R}{F_\beta}\left(\frac{f_\beta}{f_{max}}\right)^{1/2} a_{rms}$$ (11)

where a_{rms} values are taken between the S-wave pulse initiation and $1/f_\beta$. All three stress release estimates provide values on the same order of magnitude (Table 2), with σ_{rms} approximately half of $\Delta\sigma$.

Source Parameter Calculations for Inhomogeneous Faulting

Near-source Parameters

An inhomogeneous source model states that the earthquake consists of an arbitrary distribution of areas of different strength, which consequently lead to different stress drops during seismic failure. These areas can be thought of as asperities within which further asperities are assumed to exist as part of a source hierarchy. It has, however, not yet been determined to what extent self-similarity is preserved throughout the failure process. BOATWRIGHT (1988) showed that velocity and acceleration pulses are associated with asperity distributions. MCGARR (1984) associated the maximum ground motion to one of the existing asperities within the source, namely the most energetic. This asperity is of the highest concern, since it gives rise to maximum values for displacement at the source, ground velocity adjacent to the fault, and stress drop. In order to obtain estimates of these near-source parameters, a time domain analysis was carried out on rotated Canmet acceleration signals and their corresponding integrated velocities and displacements. Estimates of peak acceleration ($\underline{a}$), velocity ($\underline{v}$), and displacement ($\underline{d}$) were obtained by inspection of the vector sum of the S-wave components. This was tempered by the distance to the source, ranging from 87 to 1317 m, making phase correlation a difficult task for nearby stations. Overall, ground motion parameters, $R\underline{v}$ and $\rho R\underline{a}$, were averaged for all stations and found to have relative standard errors of 20–25%.

Assuming that the most energetic asperity is a homogeneous source within itself, estimates of asperity radius r, displacement at the source D, ground velocity adjacent to the fault $\dot{D}/2$, and dynamic stress drop $\Delta\sigma_a$ were obtained following MCGARR (1991) as

$$r = 2.34\ \beta\underline{v}/\underline{a} \qquad (12)$$

$$D = 8.06\ R\underline{v}/\beta \qquad (13)$$

$$\dot{D}/2 = 1.28\ (\beta/\mu)\rho R\underline{a} \qquad (14)$$

$$\Delta\sigma_a = 2.50\ \rho R\underline{a}. \qquad (15)$$

The constants, as defined for S waves, assume $K_\beta^B = 2.34$ (see equation (8)) and an average radiation pattern coefficient of 0.57. In deriving the above stress drop, it was considered that the asperity fails from its boundary inward so that its radius remains fixed (DAS and KOSTROV, 1983). It is worth noting that the reverse model of outward failure proposed by BOATWRIGHT (1988) leads to similar stress drop estimates. The average ground motion and calculated near-source parameters are

Table 3

Near-source parameters based on the inhomogeneous model.

No.	$R\underline{v}$ (m²/s)	$\rho R\underline{a}$ (bar)	r (m)	D (10^{-3} m)	$\dot{D}/2$ (m/s)	$\Delta\sigma_a$ (bar)
1	20.66	116	48	44	1.44	289
2	14.54	66	55	31	0.82	164
3	4.42	28	39	9	0.35	70
4	1.15	16	16	2	0.19	39
5	6.28	49	31	13	0.62	124
6	20.81	103	43	44	1.28	257
7	21.60	125	42	46	1.56	312

presented in Table 3. The $\Delta\sigma_a$ values are about 5 to 10 times larger than the $\Delta\sigma$ estimates presented in Table 2.

Reduction of South African Mining-induced Earthquakes to Canadian Data

To date, very few observational near-source studies have been reported in the literature. The study most closely related to ours, is that of MCGARR *et al.* (1981). They analyzed a series of 12 mining-induced seismic events, having local magnitudes M_L between -1 and 2.6, that were located at a depth of about 3 km in the East Rand Proprietary Mines, South Africa. For these events, band-limited acceleration signals (between 1 and 400 Hz) were recorded at several hundred meters from the source. In order to compare our results with those obtained by MCGARR *et al.* (1981), several adjustments were made to account for differences in magnitude scale (Nuttli magnitude versus local magnitude), scattering and anelastic attenuation (differences in the recording frequency band), stress regime (compressional versus extensional), and depth.

Both magnitude scales have been calibrated with seismic moment and roughly similar dependencies were found, with a shift of approximately $+0.9$ as seen in Figure 3. Since pulses at small distances were analyzed, neither study considered the effect of scattering and anelastic attenuation. However, this effect might be important since peak pulses are associated with high frequencies, particularly when comparing results at different frequencies. For S waves, assuming an average source-sensor separation of 800 m, equation (1) gives $F_a(35\text{ Hz}) = 1.6$ in our case, and $F_a(140\text{ Hz}) = 6.4$ for MCGARR *et al.*, the two frequencies chosen as average values of estimated corner frequencies using asperity radii as source radii in equation (8). Additional consideration of the stress environment, together with the differences in depth between the two studies can be expressed in a relative manner

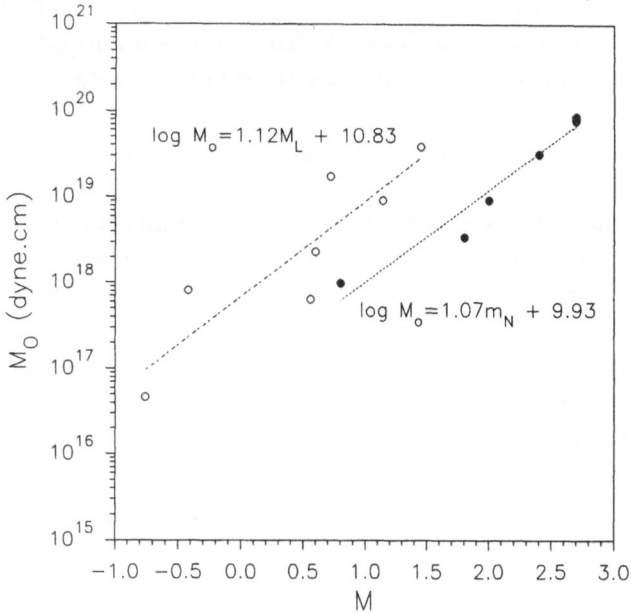

Figure 3
Locally calibrated moment-magnitude scale with M_L values (open circles) from McGARR *et al.* (1981) and m_N values (filled circles) from this study.

for the peak parameters following McGARR (1984)

$$\frac{(R\underline{v})_e}{(R\underline{v})_c} = \frac{0.07 + 0.08(\text{km}^{-1})z_e}{0.33 + 0.13(\text{km}^{-1})z_c} = 0.71$$

$$\frac{(\rho R\underline{a})_e}{(\rho R\underline{a})_c} = \frac{-10.8 + 30.6(\text{km}^{-1})z_e}{56.5 + 87.6(\text{km}^{-1})z_c} = 0.64 \qquad (16)$$

where $\rho R\underline{a}$ is in bars. The extensional regime, denoted by 'e', is at 3 km depth in their case whereas the compressional regime, denoted by 'c', is at 0.8 km depth in our case. To adjust their peak parameters to reflect our compressional domain and attenuation conditions, the following corrections were made

$$(R\underline{v})_e = \frac{F_a(140\ \text{Hz})}{F_a(35\ \text{Hz})} \cdot 0.71(R\underline{v})_c = 2.84(R\underline{v})_c,$$

$$(\rho R\underline{a})_e = \frac{F_a(140\ \text{Hz})}{F_a(35\ \text{Hz})} \cdot 0.64(\rho R\underline{a})_c = 2.56(\rho R\underline{a})_c. \qquad (17)$$

Since a linear relationship is observed between each log peak parameter and magnitude, the above differences can be equivalently expressed as an additional shift in magnitudes. The constants in equation (17) lead to magnitude shifts of -0.45 and -0.40 as related to $\log(R\underline{v})$ and $\log(\rho R\underline{a})$, respectively. Finally, by accounting for the previously determined magnitude shift due to different magnitude scales (+0.9), we can represent South African data by an overall shift in their magnitudes of +0.45 and +0.50 for log peak velocities and accelerations, respectively.

Figures 4 and 5 show the peak parameters of South African and Canadian data, as functions of magnitude, reduced to a compressional stress regime at 800 m depth. Distinct linear dependencies were observed in peak parameters with magnitude. There is, however, a large shift in $\rho R\underline{a}$, with Canadian values approximately 3 times less than South African values. This shift can be explained by differences in the recording frequency bandwidths, up to 400 Hz in their case compared to 150 Hz in our case. If we assume that the same event is recorded by two different frequency bandwidth instruments, denoted by 1 and 2, then it may be possible to theoretically estimate the expected peak parameter behavior as a function of frequency. Assuming that both instruments will provide the same seismic moment, i.e., $M_{01} = M_{02}$ (where $M_0 = \mu AD$), then $D_1 = D_2$. Since $D \propto R\underline{v} \propto \rho R\underline{a} \cdot r$, as derived from equations (12) and (13), we can consider the following

$$\frac{\Delta\sigma_{a1}}{\Delta\sigma_{a2}} = \frac{(\rho R\underline{a})_1}{(\rho R\underline{a})_2} = \frac{r_2}{r_1} = \frac{f_1}{f_2} \tag{18}$$

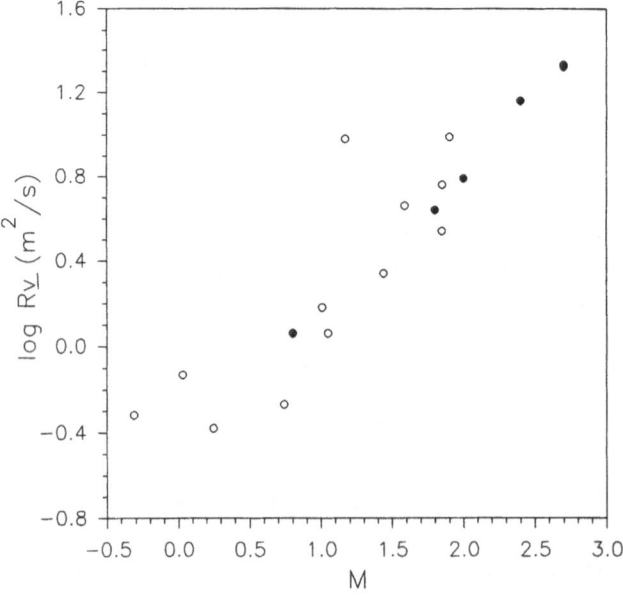

Figure 4

Peak velocity parameter as a function of magnitude; open circles represent values as modified from McGarr *et al.* (1981), whereas filled circles represent values from this study.

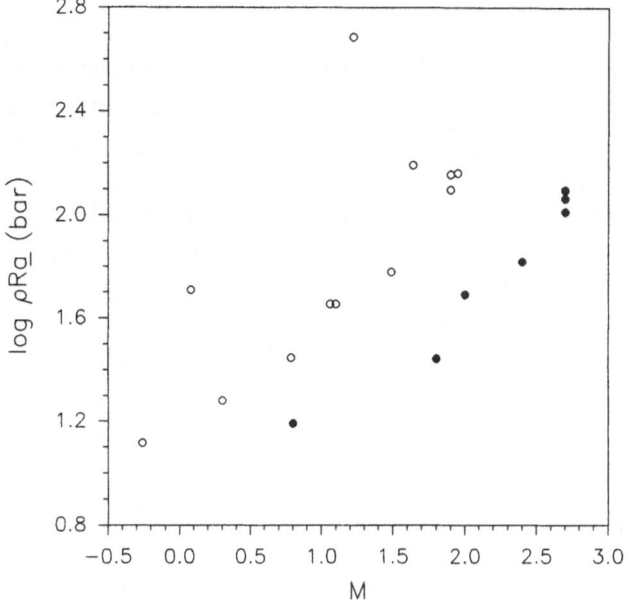

Figure 5

Peak acceleration parameter as a function of magnitude; open circles represent values as modified from MCGARR *et al.* (1981), whereas filled circles represent values from this study.

where f_1 and f_2 represents the corner frequencies, and

$$\frac{(R\underline{v})_1}{(R\underline{v})_2} = \frac{(\rho R\underline{a})r_1}{(\rho R\underline{a})r_2} = \frac{r_2}{r_1} \cdot \frac{r_1}{r_2} = 1. \tag{19}$$

As a result, a shift can be expected only in $\rho R\underline{a}$. Since accurate corner frequency determination generally requires an upper frequency bandwidth limit at least 3 to 4 times larger than the measured corner frequency (e.g., HANKS, 1977), the observed shift can then be explained by differences in the recording frequency bandwidth as suggested by the average asperity corner frequencies, $f_1/f_2 = 140/35 = 4$. It also follows from the above that no shift will exist in $R\underline{v}$, and as observed in Figure 4, no shift is seen. Based on the combined data set analysis, we can conclude that (a) highly stressed areas have a similar behavior within the inhomogeneous source, at least for frequencies up to 140 Hz; (b) absolute values of the peak near-source accelerations are highly dependent on the recording frequency range, and they appear to increase with increasing maximum frequency; and (c) that the source level (e.g., asperity dimension, dynamic stress drop) one is able to discern, is directly related to the recording instrumentation, which acts as a limiting factor.

Source Rupture Complexity

The homogeneous model assumes a simple rupture process with a constant rupture velocity and stress drop throughout the source area. This would imply that the estimates of stress release, such as static stress drop, dynamic stress, and rms stress drop would be similar. However, the inhomogeneous model inherently assumes a certain level of complexity of faulting as observed in the waveforms. This in turn suggests that different estimates of stress drop will no longer be similar, as $\Delta\sigma_a$ will be most sensitive to the stress release of the most energetic subevent, and $\Delta\sigma$ and $\Delta\sigma_{rms}$ stress drops will be influenced by the stress release from all the subevents in the faulting process. This is clearly evident when one examines the stress release ratios of $\Delta\sigma_a/\Delta\sigma$ (4.7 to 11.2) and $\Delta\sigma_a/\Delta\sigma_{rms}$ (11.9 to 21.5) as derived from Tables 2 and 3. These values allow us to conclude that a certain level of source rupture complexity exists, in agreement with estimates of r_0/r, which were found to range from 1.4 to 3.0 (Table 4).

An independent measure of source complexity can be obtained by comparing the signal duration of the slip over the entire source (τ) to the pulse duration of the rupture of the first subevent (τ_r),

$$c = \frac{\tau/\tau_r + \gamma}{1 + \gamma} \tag{20}$$

where γ is the waveform compactness. A value of $\gamma = 1$ signifies that the rupturing of existing subevents proceeds without arrest over the entire faulting surface, resulting in a compact waveform. When rupture arrests accompany individual subevents, the corresponding waveforms are characterized by smaller values of γ (BOATWRIGHT, 1984). In our approach, we used S-wave accelerations rather than velocities, however, both accelerations and velocities were checked for correct selection of signal and pulse durations. The near-field contribution at stations close

Table 4

Source complexity and rupture velocity estimates.

No.	τ/τ_r	γ	c	r_0/r	v_r
1	3.8	0.5	2.9	1.7	0.37
2	3.6	0.9	2.4	1.4	0.55
3	2.8	0.5	2.2	1.4	0.38
4	2.9	0.6	2.2	2.5	0.43
5	3.2	1.0	2.1	1.9	0.41
6	4.2	0.8	2.8	3.0	0.33
7	5.0	0.4	3.9	2.3	0.41

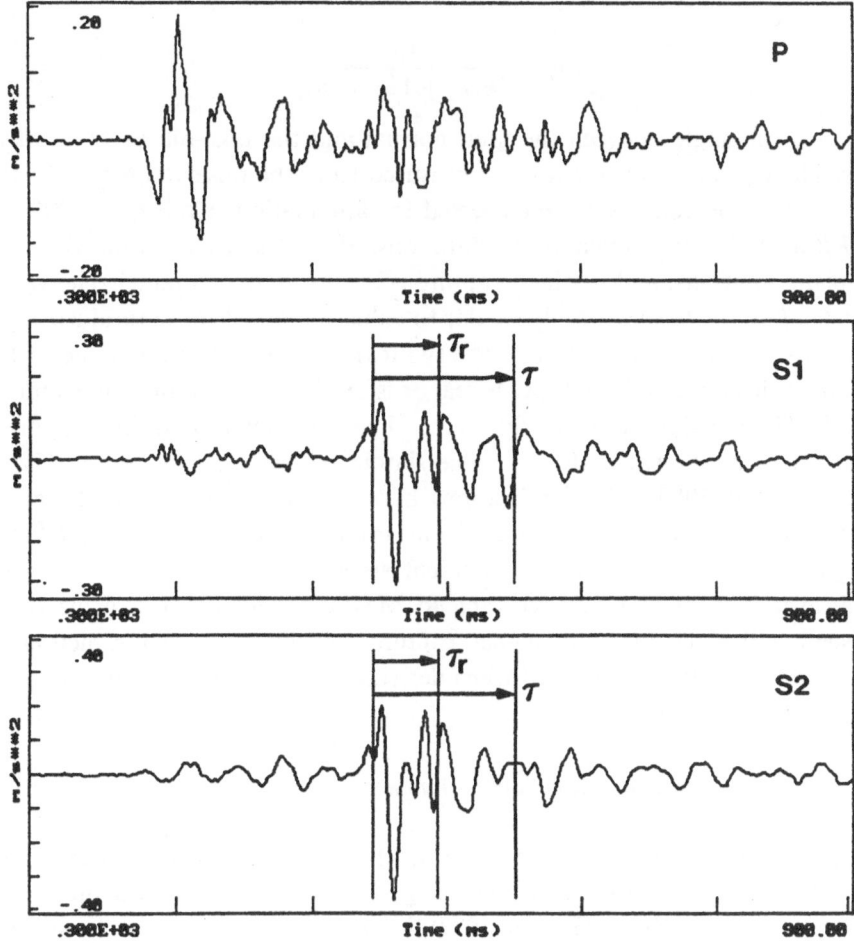

Figure 6

Example of rotated triaxial acceleration signals for event 2 at a distance of 1290 m from the source. The pulse τ_r and signal duration τ measurements are as shown.

to the source complicated the analysis, particularly when measuring the waveform compactness. Examples of signal and pulse duration measurements are given in Figure 6 and station average values are presented in Table 4.

The source complexity values ranged from 2 to 4, similar to previously derived estimates of r_0/r. As a result, r can be considered to be associated with the corresponding rupture phases and can therefore be related to the displacement pulse rise time $\tau_{1/2}$ (the delay between the first arrival and the arrival of the first stopping phase), and the rupture velocity v_r (BOATWRIGHT, 1980) by

$$v_r = \frac{13r}{16\tau_{1/2} + 12r \sin \theta/\beta} \tag{21}$$

where θ is the angle between the fault normal and the take-off direction of the S waves. The pulse rise times $\tau_{1/2}$ were measured from the observed first S-wave half cycle (τ_{obs}) of the velocity trace corrected for attenuation, $\tau_{1/2} = \tau_{obs} - t^*/2$, where $t^* = R/(\beta Q_\beta)$. Using known pulse durations, θ angles, as determined from the fault-plane solutions, and r values, rupture velocities were computed. The values lie within 0.3β to 0.6β, surprisingly similar to values obtained for natural earthquakes of small to moderate magnitude (BOATWRIGHT, 1984). We cannot, nevertheless, conclude if this observed similarity is real or is due to the limitations inherent in the method. These values are clearly lower than the 0.9β used in our previous evaluations. Consequently, the K_β^B value should be decreased by about 25% (AKI and RICHARDS, 1980). The implications of reducing K_β^B would not be dramatic since this difference is much less than the uncertainty in choosing K_β^M or K_β^B. Worth noting are the observed ratios of apparent stress to the static stress drop, which overall average around 0.5. According to MADARIAGA (1976), a value of 0.5 is consistent with a significantly higher rupture velocity. Such contradictions have been quite frequently encountered when deriving stress release estimates by different authors (SCHOLZ, 1990).

Scaling Relations and Comparison of Source Models

Scaling relations describe the manner in which the stress drop changes when the source dimension increases with increasing seismic moment. Based on numerous observational studies (e.g., HANKS, 1977), it has been found that the homogeneous source model derived stress drop is independent of source dimensions. Consequently, from (9), it can be seen that the source radius r_0 should scale as $M_0^{1/3}$. It has also been shown that, if the bulk modulus and the stress drop do not vary systematically with earthquake size, both peak ground velocity (13) and acceleration (15) parameters, as derived from Brune's model, should also scale as $M_0^{1/3}$ (McGARR, 1991). Different scaling dependence of peak velocity parameter with seismic moment, other than stated above, can be interpreted as a source effect, assuming corrections for anelastic attenuation and scattering effects have been made as previously described.

Since Figures 4 and 5 show similar dependencies of peak parameters on magnitude, and inherently on seismic moment for the two data sets considered, by rejecting outliers, the following overall fit was established

$$R\underline{v} \propto M_0^{\gamma = 0.67} \tag{22}$$

$$\rho R\underline{a} \propto M_0^{\delta = 0.48} \tag{23}$$

where the powers are mean values calculated from Canadian ($\gamma = 0.64$, $\delta = 0.44$) and South African ($\gamma = 0.69$, $\delta = 0.51$) data. McGARR et al. (1981) found that the peak velocity parameter in the far-field is consistent with Brune's model and can be expressed as

$$R\underline{v} \propto \Delta\sigma r_0 \qquad (24)$$

where $\Delta\sigma$ is the static stress drop over the entire source of radius r_0. Using relationship (24), and incorporating (9) and (22), it follows that

$$\Delta\sigma \propto M_0^{(3\gamma - 1)/2 = 0.50} \qquad (25)$$

which questions the validity of the constant stress drop assumption in creating any homogeneous type model. The inhomogeneous model (15 and 23) gives

$$\Delta\sigma_a \propto M_0^{\delta = 0.48} \qquad (26)$$

which is self-consistent with a postulated asperity distribution. It is worth mentioning that McGARR (1986) also analyzed five sequences of crustal earthquakes over a broad magnitude range ($10^{19} - 10^{24}$ dyne · cm) and found the typical γ to be 0.70. He interpreted these results as being valid below certain transition points in the seismic moment. Above this transition, the stress drop did not appear to depend on M_0, and therefore McGarr concluded that a transition in scaling exists. From all these observations, it can be suggested that different levels of asperity distribution may exist for different magnitude ranges and geophysical conditions (e.g., geological setting, stress regime), and that each level can be considered as a self-similar family showing a self-similar scaling relationship. Main shocks, which lie above transition points, would represent characteristic earthquakes and form a fractal distribution within themselves (e.g., SCHOLZ and AVILES, 1986; TRIFU and RADULIAN, 1989). In our case, a clear transition in scaling is not observed, suggesting that our events may not represent characteristic earthquakes for this magnitude domain.

The implications of selecting an inhomogeneous source model can be considered by examining the effect stress drop scaling may have on the high frequency spectral fall-off. FRANKEL (1991) assumed that for an inhomogeneous source model, characterized by a self-similar distribution of subevents or asperities, the fault strength is proportional to the dynamic stress drop. Hence, fault strength scaling with asperity size is identical to dynamic stress drop scaling ($\Delta\sigma_a$) with source dimension (r)

$$\Delta\sigma_a \propto r^\eta. \qquad (27)$$

Assuming that the subevent distribution has a fractal dimension $D_F = 2$, and subevent areas are equal to the main shock area, Frankel determined that the high-frequency fall-off χ is

$$\chi = -3 - \eta + D_F/2. \qquad (28)$$

Since the slip takes place over the entire source area ($A = \pi r_0^2$) and the stress release is associated with only the most energetic asperity, then

$$M_0 = \mu A D \propto \mu A \, \Delta\sigma_a r_a. \tag{29}$$

By substituting M_0 from equation (26) and rewriting (29) in the form of equation (27) we arrive at

$$\eta = \delta/(1 - \delta) \simeq 1. \tag{30}$$

With $\eta = 1$ and $D_F = 2$, equation (28) gives a high-frequency spectral fall-off of $\chi = -3$. Obviously, the assumptions used in deriving equation (28) may be further changed, but as it is, it requires a fall-off of ω^{-3} if stress drop or strength scaling exists ($\eta = 1$). As can be seen, it is the constant stress drop scaling ($\eta = 0$) that can only generate fall-offs of ω^{-2} for $D_F = 2$. The ω^{-3} slopes, however, are not clearly observed in the spectral analyses. This might be due to the presence of stopping phases (AKI and RICHARDS, 1980) and/or non-shear components in the failure mechanism (ARCHAMBEAU, 1968), both which can significantly enrich the high-frequency content. More recently, TUMARKIN *et al.* (1992) generalized Frankel's model by considering that the subevent distribution is not restricted to a plane ($D_F > 2$). They obtained

$$\chi = -(3 + \eta)/2 \tag{31}$$

which consequently leads to a spectral fall-off of ω^{-2} only for the inhomogeneous model when assuming the strength scaling $\eta = 1$ as derived in equation (30). This strongly favors a strength scaling relationship for the distribution of subevents at the source, rather than a constant stress drop scaling relationship.

Further to the above discussion, the displacement D, as derived from the inhomogeneous model equation (13), can be compared with direct estimates based on the near-source displacement dependence with both event magnitude and distance to the source. Although our data set is limited, there are 3 events of similar magnitudes, 2.7 ± 0.1; this allowed us to derive an attenuation relation for displacement with distance, $\underline{d} = \underline{d}(R)$. The contribution of geometrical spreading and scattering plus anelasticity can be written as

$$\underline{d}(R) \propto R^a \exp(bR) \tag{32}$$

where a has different values for the far-field ($a = -1$) and near-field ($a = -1/2$). Consequently, the peak parameter attenuation relationship at constant magnitude can be stated as follows

$$\log \underline{d}(R) = a \log R + bR + c. \tag{33}$$

In our case, for distances smaller than 460 m, anelastic attenuation is not very important ($b \simeq 0$). As a result, Figure 7 only shows the log-log dependence of

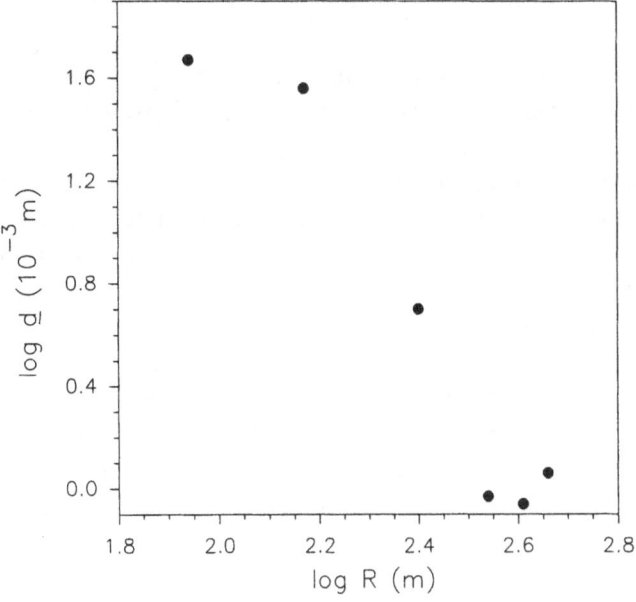

Figure 7

Relationship of maximum displacement $\underline{d}$ with distance R from the source of the magnitude 2.7 events (within 460 m).

equation (33). The tendency of ground motion to saturate near the source has been noted by many authors for a wide range of magnitudes (e.g., FUKUSHIMA and TANAKA, 1990; NIAZI and BOZORGNIA, 1991) and is expected for a limited bandwidth. By extrapolation, it follows that the displacement at the source is $\underline{d} \simeq 60$ mm, which is in fairly good agreement with the displacement as averaged from Table 3 when accounting for attenuation as discussed in the previous section, $D = 45 \times [F_a(35 \text{ Hz}) = 1.6] = 72$ mm.

Referring to the homogeneous models, the estimate of D is particularly important as it leads us to conclude that K_β^B from Brune's model provides reliable results when applied to the most energetic asperity of radius r, actually 2 times less than the overall source radius r_0. Even though the r values are in close agreement to what Madariaga's homogeneous model would estimate for the source radius, K_β^M would lead to D values over 3 times larger than actually observed. To fit these values, an exponential behavior would have to be assumed for the attenuation in close proximity to the fault. This, however, is in total disagreement with the observed attenuation relationship as shown in Figure 7. As a result, the observed ground motion scaling behavior tends to favor the use of an inhomogeneous source model.

Implications for the Evaluation of Damage Potential

The selection of a homogeneous or inhomogeneous source model has important implications for seismic hazard assessment (prediction of ground motion level) of underground facilities, such as mines, waste disposal repositories, and underground research laboratories. For example, the issue of seismic hazard assessment at Creighton mine in the Sudbury Basin has been of immediate concern as it is the siting of the Sudbury Neutrino Observatory (WOLFENSTEIN and BEIER, 1989). Extensive work has been carried out by KAISER *et al.* (1992) at Creighton mine, where geomechanical techniques have been applied to quantify the damage level of both the rock mass and the structural support resulting from rockbursts (the sudden failure of highly stressed rock in the vicinity of openings, leading to the expulsion of rock). They developed an empirical scale based on macroseismic observations, equivalent to the earthquake intensity scale. This scale has five levels and using blast data they derived an empirical formula to relate these levels to peak particle velocities

$$\underline{v} = 50 \times 2^{(DL-1)} \tag{34}$$

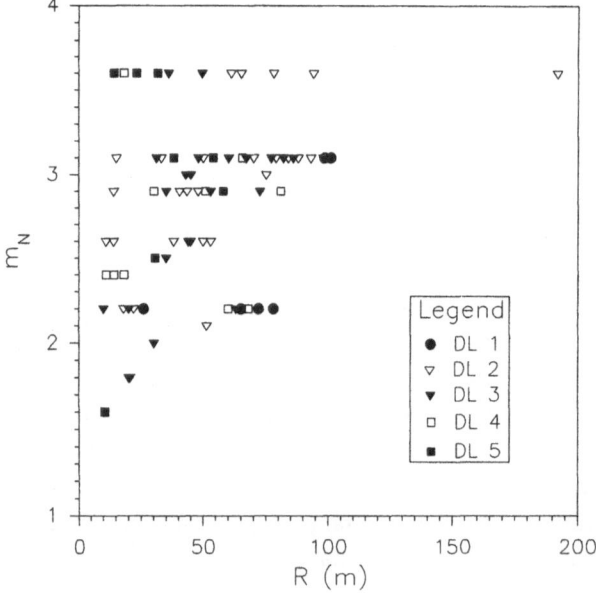

Figure 8

Measurements of rock mass damage as a function of rockburst magnitude and distance to the hypocenter, as modified from KAISER *et al.* (1992), where *DL* defines zones of different damage levels related to peak particle velocities by equation (34).

where *DL* is the observed damage level and peak particle velocities are in mm/s. Figure 8 shows estimated damage levels based on geomechanical measurements at Creighton mine as a function of magnitude and distance to the source.

Strathcona mine lies within the same stress province as Creighton mine and within the same geologic domain, suggesting that a direct comparison of our results can be made with the above data. Table 5 gives the average distances from the source origin of the 2.7 magnitude events in this study, as determined from peak velocity parameters in Table 3, and from Figure 8 as a function of damage level. The corresponding peak particle velocities, as derived from equation (34), are also provided in Table 5. Despite the differences in methodologies, there is very good agreement between the distance estimates at which higher damage levels (3 to 5) would occur. However, there are significant differences between the distance estimates correponding to lower damage levels (1 and 2), with our estimates about twice those based on the underground geometrical data. This disagreement may be explained by the use of blast data to calibrate the damage scale employed in Figure 8. Blasts contain higher frequencies than expected for a shear failure mechanism. The corresponding ground motions at these higher frequencies will tend to attenuate faster with distance from the source, resulting in smaller distance estimates at which damage would occur. These differences are accentuated at lower damage levels, corresponding to relatively larger distances, for which the attenuation effects are expected to be more significant. In spite of these discrepancies, the similarities at smaller distances provide unique evidence supporting our prediction of the ground motion level based on the inhomogeneous model.

Following the discussion above, it is worth noting that the region of serious damage, delineated by damage levels 3 to 4, and our asperity radii estimates are closely correlated. As such, we believe that the radii over which damage occurs are directly related to the asperity radii rather than to the overall source radii as

Table 5

The distance from the source and the damage level (DL) attained for a magnitude 2.7 earthquake based on (a) near-source strong motion records, and (b) geomechanical measurements in the Creighton mine. Peak particle velocities $\underline{v}$ corresponding to the damage levels are also provided.

DL	$\underline{v}$ (mm/s)	Distance[a] (m)	Distance[b] (m)
1	50	400	170
2	100	200	120
3	200	100	80
4	400	50	55
5	800	25	35

predicted by homogeneous models. In fact, due to the frequency bandwidth used, our asperity measurements suggest an upper limit of the highest stress release region. Further to this, the measured displacement at the source leads us to consider the inhomogeneous model as being the most appropriate for damage potential assessment. Through a homogeneous model, only K_β^M leads to small enough source radii to be comparable to the observed damage area, but this gives unrealistic D values. However, when using the inhomogeneous model, K_β^B gives both realistic source radii and displacement values.

Conclusions

In this study, we have examined homogeneous and inhomogeneous model derived source parameters of seven mine-induced events ranging in magnitude from $m_N 0.8$ to 2.7 located between 640 and 825 m depth at Strathcona mine, Sudbury, Ontario. The analysis included an investigation of peak velocity and acceleration scaling relations with magnitude for available near-source Canadian and South African data. Additional source rupture complexity measurements were made to better constrain the selection of a source model to be used in the evaluation of damage potential. Finally, to validate the use of this source model for seismic hazard assessment, a comparison of its predicted ground motion level with underground geomechanical measurements was carried out.

From spectral analysis of rotated waveforms, estimates of seismic moment, seismic energy, source radii, and stress release (static stress drop, apparent stress, and rms stress drop) were obtained. Results based on the Brune source model indicate that the static stress drop is similar to that observed for natural earthquakes, on the order of 30 bars. All three stress release parameters had values on the same order of magnitude. However, Madariaga's model derived source radii better fit documented evidence of underground damage, suggesting that static stress drops should be 5.57 times larger than estimated by Brune's model. Worth noting is an apparent 2 to 4 times larger contribution of the P waves to the total energy than expected for a pure-shear failure mechanism. Regardless, the fit of the double-couple model to the distribution of P-wave polarities indicates that shear is the dominant mode of failure.

For the inhomogeneous source model, estimates of asperity radius, slip, slip velocity, and dynamic stress drop were obtained. In comparison to other results (McGARR *et al.*, 1981), we showed that the absolute value of the peak acceleration, and consequently the asperity 'level' one is able to discern is directly dependent on the recorded frequency content. The asperity radii are on average 2 times smaller than the overall source radii based on Brune's model. The displacement attenuation relationship with distance to the source was found to saturate in the source proximity ($\leq 2r$) at about 60 mm for magnitude 2.7 events, in good agreement with slip values derived when using peak velocity measurements and assuming the asperity is a homogeneous Brune source within itself (72 mm). Note, if the asperity is considered as a homogeneous Madariaga source, the displacement at the source is about 3 times larger than actually determined.

In terms of the inhomogeneous model, a stress drop scaling was found, $\Delta\sigma_a \propto M_0^{1/2}$, similar to that obtained by MCGARR (1986) for crustal earthquakes with $M_0 = 10^{19}-10^{24}$ dyne · cm. This behavior suggests that relative to overall source dimension, the size of the asperities responsible for the peak ground motion follows distributions that may be similar over a wide range of magnitudes and hierarchical source levels. Assuming that the fault strength is proportional to the stress drop, a scaling of the strength with asperity dimension as $\Delta\sigma_a \propto r$ is suggested. We also showed that for a self-similar planar distribution of asperities, the above leads to a high-frequency spectral fall-off of ω^{-3}. However, this tendency may be difficult to observe on real waveforms when strong stopping phases or non-shear components of failure are generated. Regardless, in assuming a volumetric distribution of asperities, a ω^{-2} fall-off can be retrieved (TUMARKIN et al., 1992).

Based on measurements of source rupture complexity (ranging from 2 to 3), which were in close agreement to expected estimates from the ratio r_0/r, the sources could be considered to be somewhat complex. Additionally, the complexity values and rupture velocities ($0.3\,\beta$ to $0.6\,\beta$) appeared to be similar to values obtained for natural earthquakes (BOATWRIGHT, 1984), further supporting the idea of a complex source and rupture process. This was also evident in the relatively large $\Delta\sigma_a$ values, as compared to the stress release estimates based on the homogeneous source model, suggesting that failure is likely associated with the most energetic subevent rather than the entire source. These observations strongly encourage the use of an inhomogeneous source model in any evaluation of source parameters.

Validation of the inhomogeneous source model appropriateness was achieved by direct comparison of the predicted ground motion level to observed damage. For distances close to the source (< 100 m), corresponding to relatively higher damage levels, a good agreement was found between the predicted peak particle velocities for the inhomogeneous model and velocities derived based on established geomechanical relationships. Away from the source, significant differences exist between these two approaches, due to the use of blast data (different attenuation behavior), in the derivation of the geomechanical relations. The high degree of similarity between the asperity radii and the regions of the highest observed local damage gives additional support for the use of the inhomogeneous source model in the assessment of seismic hazard. Overall, we conclude that the inhomogeneous source model provides reliable ground motion predictions and should be employed when considering the placement of underground facilities.

Acknowledgments

We are grateful to Canmet Laboratories for providing their data and Falconbridge Ltd. for their financial support in the integrated analysis of the rockburst sequence at Strathcona mine. We thank the personnel at Strathcona mine for their cooperation and support in this project. The authors would also like to thank the Mining Research Directorate of Canada and the Natural Sciences and Engineering Council of Canada for their financial contributions to this project. We thank Art McGarr and Ralph Archuleta for some useful suggestions and Art Lerner-Lam for his insightful review of the manuscript.

REFERENCES

AKI, K., and RICHARDS, P. G., *Quantitative Seismology: Theory and Methods* (W. H. Freeman, San Francisco, Calif. 1980).

ANDREWS, D. T., *Objective determination of source parameters and similarity of earthquakes of different size*. In *Earthquake Source Mechanics* (Das, S., Boatwright, J., and Scholz, C. H., eds.) American Geophysical Union Monograph *37* (AGU, Washington, DC 1986) pp. 259–267.

ARCHAMBEAU, C. B., (1968), *General Theory of Elastodynamic Source Fields*, Rev. Geophys. *6*, 241–288.

BOATWRIGHT, J. (1980), *A Spectral Theory for Circular Seismic Sources; Simple Estimates of Source Dimension, Dynamic Stress Drop, and Radiated Seismic Energy*, Bull. Seismol. Soc. Am. *70*, 1–27.

BOATWRIGHT, J. (1984), *The Effect of Rupture Complexity on Estimates of Source Size*, J. Geophys. Res. *89*, 1132–1146.

BOATWRIGHT, J. (1988), *The Seismic Radiation from Composite Models of Faulting*, Bull. Seismol. Soc. Am. *78*, 489–508.

BOATWRIGHT, J., and FLETCHER, J. B. (1984), *The Partition of Radiated Energy between P and S Waves*, Bull. Seismol. Soc. Am. *74*, 361–376.

BRUNE, J. N. (1970), *Tectonic Stress and the Spectra of Seismic Shear Waves from Earthquakes*, J. Geophys. Res. *75*, 4997–5009.

BRUNE, J. N. (1971), *Correction*, J. Geophys. Res. *76*, 5002.

BURDICK, L. J. (1978), *t* for S Waves with Continental Ray Path*, Bull. Seismol. Soc. Am. *68*, 1013–1030.

DAS, S., and KOSTROV, B. (1983), *Breaking of a Single Asperity: Rupture Process and Seismic Radiation*, J. Geophys. Res. *88*, 4277–4288.

FRANKEL, A. (1991), *High-frequency Spectral Fall-off of Earthquakes, Fractal Dimension of Complex Rupture, b Value, and the Scaling of Strength on Faults*, J. Geophys. Res. *96*, 6291–6302.

FUKUSHIMA, Y., and TANAKA, T. (1990), *A New Attenuation Relationship for Peak Horizontal Acceleration of Strong Ground Motion in Japan*, Bull. Seismol. Soc. Am. *80*, 757–783.

GENDZWILL, D. J., and PRUGGER, A. F., *Algorithms for micro-earthquake locations*, Proc. 4th Symp. on *Acoustic Emissions and Microseismicity* (Pennsylvania State University, University Park, Pennsylvania 1985) pp. 601–615.

GIBOWICZ, S. J., *The Mechanism of Large Mining Tremors in Poland*. In *Rockbursts and Seismicity in Mines* (Gay, N. C., and Wainwright, E. H., eds.) (South African Inst. Min. Metall., Johannesburg 1984) Symp. Ser. *6*, pp. 17–28.

GIBOWICZ, S. J., HARJES, H. P., and SCHAFER, M. (1990), *Source Parameters of Seismic Events at Heinrich Robert Mine, Ruhr Basin, Federal Republic of Germany: Evidence for Non-double-couple Events*, Bull. Seismol. Soc. Am. *80*, 1–22.

HANKS, T. C. (1977), *Earthquake Stress Drops, Ambient Tectonic Stresses, and Stresses that Drive Plates*, Pure and Appl. Geophys. *115*, 441–458.

HANKS, T. C. (1982), f_{max}, Bull. Seismol. Soc. Am. *72*, 1867–1879.

HANKS, T. C., and MCGUIRE, R. K. (1981), *The Character of High-frequency Strong Ground Motion*, Bull. Seismol. Soc. Am. *71*, 2071–2095.

KAISER, P. K., JESENAK, P., MCCREATH, D. R., and TENNANT, D. D. (1992), *Rockburst Damage-potential Assessment*, Geomechanics Research Centre, Laurentian University, Sudbury, Ontario, Canada.

MADARIAGA, R. (1976), *Dynamics of an Expanding Circular Fault*, Bull. Seismol. Soc. Am. *66*, 639–666.

MAKUCH, A., *Design of a New Macroseismic Monitoring System* (Canmet Special Publication SP86-14E) (Canadian Government Publishing Centre, Ottawa 1986), 19 pp.

MATSUMURA, S. (1981), *Three-dimensional Expression of Seismic Particle Motions by the Trajectory Ellipsoid and its Application to the Seismic Data Observed in the Kanto District, Japan*, J. Phys. Earth *29*, 221–239.

MCGARR, A. (1981), *Analysis of Peak Ground Motion in Terms of a Model of Inhomogeneous Faulting*, J. Geophys. Res. *86*, 3901–3912.

MCGARR, A. (1984), *Scaling of Ground Motion Parameters, State of Stress, and Focal Depth*, J. Geophys. Res. *89*, 6969–6979.

MCGARR, A., *Some observations indicating complications in the nature of earthquake scaling*. In *Earthquake Source Mechanics* (Das, S., Boatwright, J., and Scholz, C. H., eds.) American Geophysical Union Monograph *37* (AGU, Washington, DC 1986) pp. 217–225.

MCGARR, A. (1991), *Observations Constraining Near-source Ground Motion Estimated from Locally Recorded Seismograms*, J. Geophys. Res. *96*, 16495–16508.

MCGARR, A., GREEN, R. W., and SPOTTISWOODE, S. M. (1981), *Strong Ground Motion of Mine Tremors: Some Implications for Near-source Ground Parameters*, Bull. Seismol. Soc. Am. *71*, 295–319.

NIAZI, M., and BOZORGNIA, Y. (1991), *Behavior of Near-source Peak Horizontal and Vertical Ground Motions over SMART-1 Array, Taiwan*, Bull. Seismol. Soc. Am. *81*, 715–732.

NUTTLI, O. N. (1973), *Seismic Wave Attenuation and Magnitude Relations for Eastern North America*, J. Geophys. Res. *78*, 876–885.

SCHOLZ, C. H., *The Mechanics of Earthquake and Faulting* (Cambridge University Press, New York 1990).

SCHOLZ, C. H., and AVILES, C. A., *The fractal geometry of faults and faulting*. In *Earthquake Source Mechanics* (Das. S., Boatwright, J., and Scholz, C. H., eds.) American Geophysical Union Monograph *37*. (AGU, Washington, DC 1986) pp. 147–155.

SNOKE, J. A. (1987), *Stable determination of (Brune) Stress-drops*, Bull. Seismol. Soc. Am. *77*, 530–538.

THURBER, C. H. (1985), *Nonlinear Earthquake Location: Theory and Examples*, Bull. Seismol. Soc. Am. *75*, 779–790.

TRIFU, C.-I., and RADULIAN, M. (1989), *Asperity Distribution and Percolation as Fundamentals of an Earthquake Cycle*, Phys. Earth Planet. Interiors *58*, 277–288.

TUMARKIN, A. G., ARCHULETA, R. J., and MADARIAGA, R. (1992), *Basic Scaling Relations for Composite Earthquake Models*, Abstract, EOS Trans. Amer. Geophys. Union *73*, 389.

URBANCIC, T. I., and YOUNG, R. P. (1990), *Focal Mechanism and Source Parameter Studies of a m_N 1.8–2.7 Sequence of Mining-induced Seismic Events Recorded during June, 1988, at the Strathcona Mine, Sudbury, Canada*, Tech. Trans. Project Rep., Dept. Geol. Sci., Queen's Univ., Kingston, Canada.

WOLFENSTEIN, L., and BEIER, E. (1989), *Neutrino Oscillations and Solar Neutrinos*, Phys. Today *42* (7), 28–36.

(Received June 24, 1994, revised/accepted February 8, 1995)

PAGEOPH, Vol. 145, No. 1 (1995)

0033–4553/95/010029–09$1.50 + 0.20/0

Induced Seismicity in the Khibiny Massif (Kola Peninsula)

Elena O. Kremenetskaya[1] and Victor M. Trjapitsin[1]

Abstract —The topic of this paper is to review recent processes of increasing seismic activity in the Khibiny Massif in the Kola Peninsula. It is a typical example of induced seismicity caused by rock deformation due to the extraction of more than $2 \cdot 10^9$ tons of rock mass since the mid-1960s. The dependence of seismic activity on the amount of extracted ore is demonstrated. Some of the induced earthquakes coincide with large mining explosions, thus indicating a trigger mechanism. The largest earthquake, which occurred on 16 April 1989 ($M_L = 4.1$) could be traced along the surface for 1200 m and observed to a depth of at least 220 m. The maximum measured displacement was 15–20 cm.

Key words: Rock excavation, mining explosion, induced seismicity.

Introduction

The contemporary seismicity in northern Europe is shown in Figure 1. Most of the earthquakes shown are for the past 30-years period, but some historical epicenters are also included, dating back as far as the year 1542. The earthquake activity is characterized by small and moderate size earthquakes. The largest earthquake in the Baltic Shield in recent years was the $M_L = 5.2$ event of 20 May 1967, located at 66.6°N, 33.7°E (MEYER and AHJOS, 1985). It was felt with MSK intensity VII in Karelia, and generated a small tsunami in the White Sea.

The Khibiny alkaline Massif (see Fig. 2) situated in the middle of the Kola Peninsula is tectonically unstable. Its seismic activity has been characterized by groups of a few earthquakes with typically 8–10 years between groups (PANASENKO, 1969). The Massif consists of different blocks separated from each other by faults. The recent crustal movements are in the range of 2–4 mm/year (YAKOVLEV, 1982).

Due to a great variety of valuable minerals found in the Kola Peninsula, the mining industry has been extensively developed and is now the main base for the economy in the Murmansk region. A majority of towns and settlements are based on giant mineral processing companies. Examples include Apatity and Kirovsk

---

[1] Kola Regional Seismological Centre, Apatity, Russia.

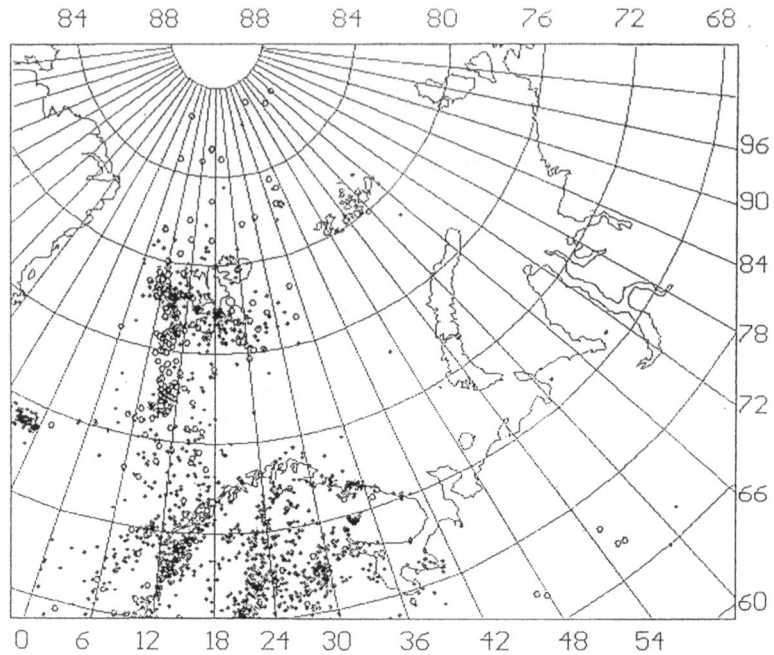

Figure 1
Epicentral map of earthquakes in the European Arctic region. Most of earthquakes shown are from the period 1960–1990, but some earlier events have been included. $+$, $m < 4.0$; $\bigcirc$, $m \geq 4.0$.

(apatite-nepheline); Monchegorsk, Nikel, Zapolyarnyi and Pechenga (nickel and copper); Kovdor, Yona, Rikolatva and Olenegorsk (mica); and Revda (rare metals).

It is known that an increase in seismicity in seismic areas and the generation of seismicity in aseismic areas have been observed as a result of deep underground mining and large-scale surface quarrying, the injection of fluids in rocks at depth, the removal of fluids from subsurface formation, and the detonation of large underground explosions (MCGARR, 1991; GIBOWICZ, 1990).

The exploitation of the Khibiny apatite ores started in 1929, and since then about $2.5 \cdot 10^9$ tons of rock have been mined from an area of about 10 km^2. This corresponds to a decrease of the gravitational component by typically 2.5–3.0 MPa, and for some parts of the Massif by as much as 9–12 MPa.

At the present time more than 10^8 tons of ore are extracted annually from three underground and three open-pit mines. The velocity of the uplift of the near-surface

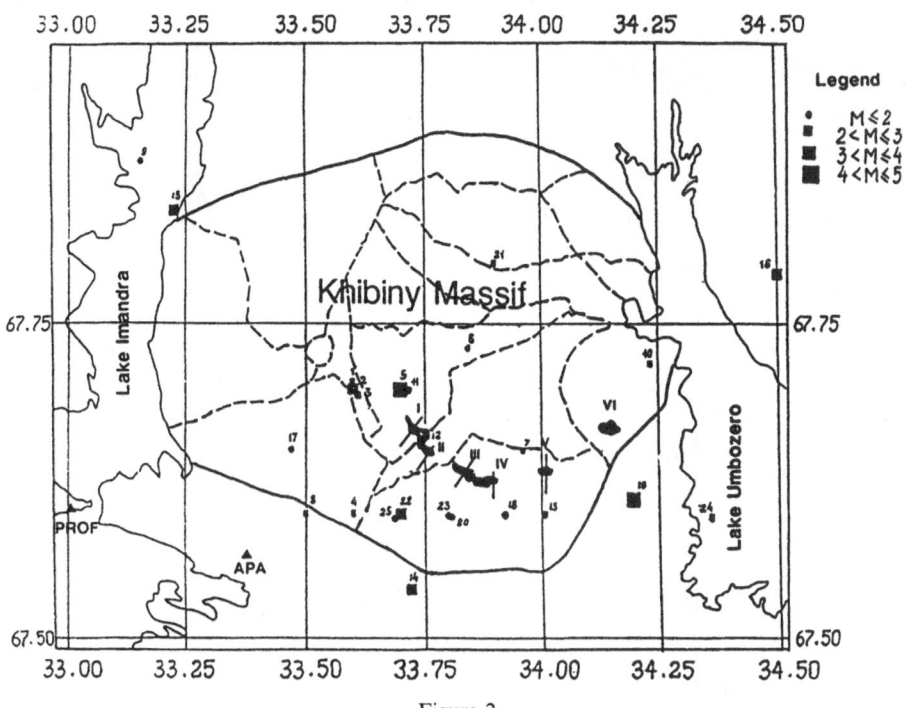

Figure 2

The position of mines (I–VI) in the Khibiny Massif together with fault structures (stippled lines) and earthquakes. Mines I, II and III are underground, whereas IV, V and VI are open-pit mines. The location of the Apatity seismic station (APA) is also shown.

parts is of the order of 70 mm/year for some tunnels (PANASENKO and YAKOVLEV, 1983).

Today, the Khibiny Massif is a typical example of increasing seismic activity caused by rock deformation due to the extraction of 100 million tons of rock mass per year. There is also evidence that some of the induced earthquakes are triggered by large mining explosions.

Seismic Activity of the Khibiny Massif

Seismicity induced by mining is usually defined as the appearance of seismic events caused by rock failures as a result of changes in the stress field in the rockmass near mining excavations (COOK, 1976). Measurements of the lithostatic stress made inside mines (MARKOV, 1977) have shown that the values of the stress horizontal components are about 30–80 MPa for the Khibiny Massif. The extraction of rockmass has led to a decrease of the vertical components for some parts of the Massif by 9–12 MPa (PANASENKO, 1983).

The extensive mining has disrupted the natural geodynamic process in the area, causing a redistribution of crustal stress, which in turn has led to increased seismic

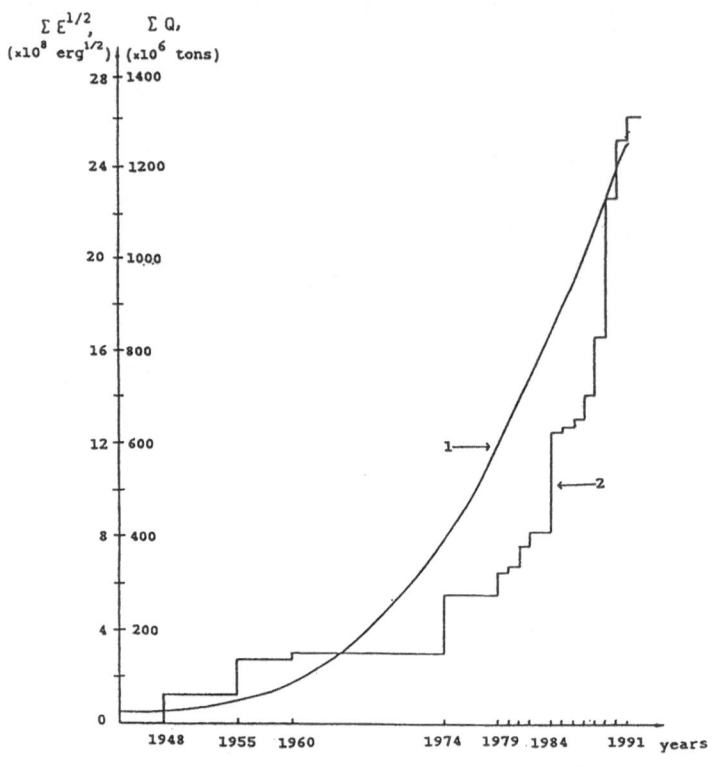

Figure 3

Cumulative distributions versus time of extracted masses of the apatite mineral (1) and Benioff's graph of released energy (2) for the Khibiny Massif. Note the similarity between the two graphs.

activity. Figure 3 illustrates in this respect the cumulative seismic energy release and the amount of extracted ore for the mines in the Khibiny Massif. The seismic energy (E) released there has been calculated from the formula (MEYER and AHJOS, 1985):

$$\log E = 12.30 + 1.27\, M_L \tag{1}$$

where E is expressed in ergs. The similarity of the two curves strongly indicates a causal connection. It should be noted here that the earthquake catalogue is considered homogeneous dating to 1978.

The dependence of seismic activity and the extracted deposit volume has long been known from observations (GLOWACKA and KIJKO, 1989). In the Khibiny Massif a more intensive excavation began in the mid-1960s, while the first significant tremors occurred in 1981. During that year four felt earthquakes (intensity $I = III-IV$ on the MSK scale) occurred in the vicinity of the mines. Some of these earthquakes were accompanied by sonic effects.

Around the same time, many rockbursts occurred near mines I and II (see Fig. 2), each of them displacing tons of rocks ($1-10\ m^3$). In fact, during five hours on

17 May 1981, more than 20 rockbursts occurred. And for the first time the occurrence of a new earthquake at the time of an explosion was observed. This situation is now typical for events in the Khibiny Massif. A possible mechanism for the blast-triggered earthquakes may be changes in the static stress induced by blasting (REASENBURG *et al.*, 1992).

The Largest Earthquakes

An earthquake with intensity more than V in the nearby town of Kirovsk occurred on 29 August, 1982, immediately after an explosion of 106 tons in mine II and at the time of a smaller quarry explosion (4.6 tons) between mines I and II.

The next significant earthquake, felt with intensity $I = V$ at Kirovsk, was on 19 June, 1984. Less than one second before this earthquake there was a small explosion (40 kg, mine II). A large block of rock ($65 \times 70 \times 70$ m^3) was broken and one part of it was displaced relative to another by about 5 cm. After this earthquake there were so many rockbursts that it was impossible to carry out work in this mine for two weeks.

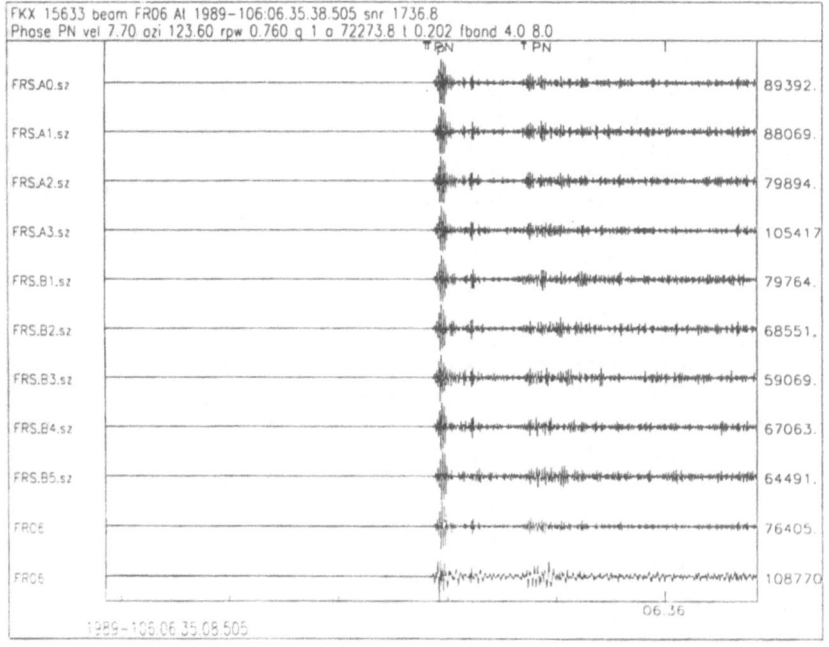

Figure 4
ARCESS recordings of the 16 April 1989 earthquake ($M_L = 4.1$) in the Khibiny Massif. The earthquake occurred almost simultaneously with a 240 ton explosion, and it is not possible to visually separate the *P* phases for the two events.

Table 1

Known earthquakes ($M_L > 2.0$) in the Khibiny Massif 1948–1991. The distance of the APA station in Apatity is indicated

No.	Date	Origin time	Distance (km)	Coordinates		ML	Seismic energy (ergs)
1	1948	0923 0000	17.0	67.70N	33.60E	3.1	1.58E16
2	1955	0808 172059	17.0	67.70N	33.60E	3.2	2.24E16
3	1955	0831 2115	17.0	67.70N	33.60E	2.5	2.80E15
4	1960	0209 210731	10.0	67.60N	33.60E	2.0	7.08E14
5	1974	0930 091142	17.0	67.70N	33.70E	3.5	6.31E16
6	1979	1212 113452	13.0			2.1	9.27E14
7	1981	0416 205634	20.0			2.1	9.27E14
8	1981	0517 075528	17.0			2.3	1.30E15
9	1981	0818 000747	19.2			2.5	2.98E15
10	1982	0422 110258	22.5			2.1	9.27E14
11	1982	0829 053335	17.0	67.70N	33.70E	3.3	3.10E16
12	1984	0619 054731	17.0	67.33N	33.70E	3.9	1.79E17
13	1984	1030 105158	35.0			2.2	1.24E15
14	1984	1030 142148	17.5	67.68N	33.72E	2.3	1.66E15
15	1987	0725 161339	22.5	67.66N	33.90E	2.9	9.60E15
16	1988	0113 025153	25.0	67.73N	33.83E	2.6	4.00E15
17	1988	0118 020948	24.0	67.65N	33.96E	2.6	4.00E15
18	1988	0120 121510	6.0	67.60N	33.50E	2.6	4.00E15
19	1988	0211 124113	27.0			2.2	1.20E15
20	1988	0304 231701	18.0	67.70N	33.70E	2.1	9.30E14
21	1988	0416 115725	18.0	67.66N	33.75E	2.1	9.30E14
22	1988	0622 013408	10.5	67.65N	33.47E	2.4	2.20E15
23	1988	1006 094741	32.0	67.61N	34.19E	3.3	3.10E16
24	1988	1123 211108	23.0	67.60N	33.80E	2.5	3.00E15
25	1989	0203 102741	33.7	67.80N	33.90E	2.2	1.20E15
26	1989	0416 063442	18.0	67.61N	33.81E	4.1	3.20E17
27	1989	0707 114924	26.5	67.71N	33.93E	3.4	4.10E16
28	1989	0724 223234	15.0	67.60N	33.78E	2.5	3.00E15
29	1989	0804 012618	21.0	67.60N	33.90E	2.1	9.30E14
30	1990	0210 163907	37.1	67.89N	33.48E	2.2	1.20E15
31	1990	0403 075417	26.5	67.60N	33.90E	2.1	9.30E14
32	1990	0612 113532	36.7	67.63N	34.29E	2.1	9.30E14
33	1990	0621 130954	20.5	67.67N	33.83E	2.3	1.70E15
34	1990	0621 231904	15.5	67.67N	33.66E	2.2	1.20E15
35	1990	0622 015115	16.5	67.61N	33.81E	2.4	2.20E15
36	1990	0624 065848	19.4	67.61N	33.88E	2.4	2.20E15
37	1990	0625 062215	16.2	67.56N	33.682	2.3	1.70E15
38	1990	0625 113116	16.1	67.61N	33.80E	2.2	1.20E15
39	1990	0625 215128	15.3	67.67N	33.66E	2.3	1.70E15
40	1990	0626 052403	19.1	67.72N	33.30E	2.4	2.20E15
41	1990	0627 004129	24.4	67.77N	33.29E	2.3	1.70E15
42	1990	0627 025428	19.6	67.67N	33.79E	2.2	1.20E15
43	1990	0627 051520	23.6	67.59N	33.99E	2.4	2.20E15
44	1990	0627 064141	29.0	67.66N	34.07E	2.8	7.20E15
45	1990	0628 015919	22.8	67.74N	33.22E	2.6	4.00E15
46	1990	0629 041211	19.4	67.69N	33.75E	2.8	7.20E15

Table 1(*Contd*)

No.	Date	Origin time	Distance (km)	Coordinates		ML	Seismic energy (ergs)
47	1990	0630 045139	20.4	67.62N	33.90E	2.4	2.20E15
48	1990	0630 064224	28.0	67.64N	34.10E	2.6	4.00E15
49	1991	0305 000836	23.7	67.55N	34.00E	2.1	9.30E14
50	1991	0505 054802	28.5	67.65N	34.07E	2.1	9.30E14
51	1991	0815 083545	24.1	67.63N	33.98E	2.1	9.30E14
52	1991	1229 083545	27.0	67.70N	33.96E	2.1	9.30E14
53	1991	1229 152044	25.0	67.57N	34.03E	2.1	9.30E14

The strongest such earthquake ($M_L = 4.1$), on 16 April 1989, occurred almost simultaneously with a large explosion (240 tons) in mine I. This earthquake was felt with I = VIII in the upper levels of the mine and I = V–VI at Kirovsk. The maximum measured displacement was 20 cm, and it occurred along a fault striking at 125–135° and dipping at 30–35°NE. The displacement was traced along the surface for 1200 m and observed to a depth of at least 220 m. This event was recorded by all seismic stations within a distance of 1000 km. Figure 4 shows record-

Figure 5
A photograph showing effects of the earthquake on 16 April 1989 inside one of the Khibiny mines. The railroad track was straight before the earthquake occurred.

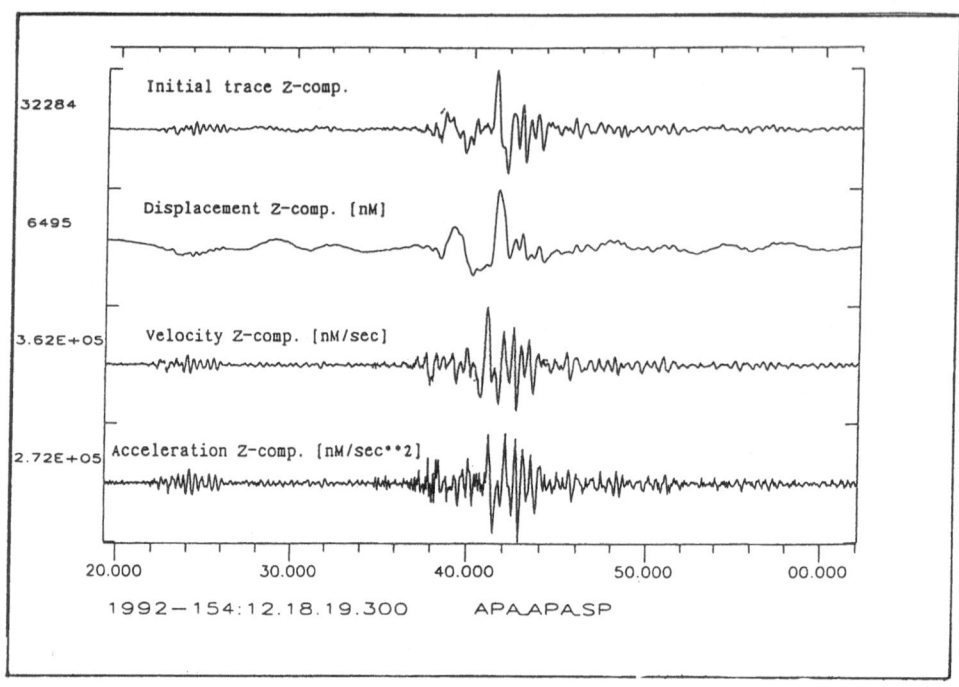

Figure 6

A double earthquake in the Khibiny Massif on 2 June 1992 (M_L(ARCESS) = 2.0 and 2.5, respectively). The figure shows APA SPZ recordings, initial trace as well as displacement, velocity and acceleration. These earthquakes were not associated with any mining explosion.

ings of this earthquake (distance 420 km) by the ARCESS array in northern Norway.

The earthquake was accompanied by a great number of aftershocks, attaining several hundred for the two subsequent months. On 24 July 1989, another earthquake occurred here. Its size was considerably smaller (see Table 1), however, remarkable destruction again took place in this mine.

At present it is too early to make a definite conclusion about the triggering mechanism between explosions and earthquakes. In some cases pairs of such events are separated by milliseconds (for example, 16 April 1989, Fig. 4); in other cases they are separated by seconds or more, sometimes they are not connected in time (for example, 2 June 1992, Fig. 6). Figure 5 is a photograph showing visible effects from the 16 April 1989, earthquake. It can be seen that the ground movement caused the railroad track inside the mine to be bent significantly.

Discussion and Conclusions

This paper had discussed two principal topics: a) the question of induced seismicity caused by rock excavation and b) the question of mining explosions as a trigger for earthquakes and rockbursts.

It has been impossible to study in detail the spatial association of pairs of explosions and triggered earthquakes. The reason is that the precision in the location of the earthquake hypocenters has so far not been sufficiently accurate. However, it appears that the induced earthquake most often occurs in the same mine in which explosion took place.

The mechanism for explosion-triggered earthquakes is at present unknown. A possible mechanism may be the changes in static stress induced by blasting. In this regard, several papers, particularly after the 1992 Landers, California, earthquake, discuss how stress release on one fault may trigger seismicity on nearby faults (STEIN et al., 1992; REASENBURG and SIMPSON, 1992).

In conclusion, the removal of large volumes of rock mass in the Khibiny Massif has created an isostatic imbalance in the crust. Although the triggering mechanisms are unknown, the deformational response to this imbalance is a significant and continuing increase in seismic activity as demonstrated in this paper.

REFERENCES

COOK, N. G. W. (1976), Seismicity Associated with Mining, Eng. Geol. (Amsterdam) 10, 99–122.

GIBOWICZ, S. J., Seismicity induced by mining. In Advances in Geophysics 32 (Dmowska, R. and Saltzman, B. eds.) (Academic Press, San Diego 1990).

GLOWACKA, E., and KIJKO, A. (1989), Continuous evaluation of seismic hazard induced by the deposit extraction in selected mines in Poland, ESC Proceedings, 21st Gen. Ass. Sofia, Bulgaria, 23–27 August 1988, 444–451.

MARKOV, G. A. (1977), Tectonic Stress and Rock Pressure in Khibiny Mines, Leningrad, Nauka, 187 pp. (in Russian).

MCGARR, A. (1991), On a Possible Connection between Three Major Earthquakes in California and Oil Production, Bull. Seismol. Soc. Am. 81, 948–970.

MEYER, K., and AHJOS, T. (1985), Temporal Variations of Energy Release by Earthquakes in the Baltic Shield, Geophysica 21 (1), 51–64.

PANASENKO, G. D. (1969), Seismic Features of the Northeast Baltic Shield, Leningrad, Nauka, 184 pp. (in Russian).

PANASENKO, G. D., Technogenic increase of tectonic activity in the Khibiny Massif, problems and ways to study it. In Geophysical Investigations in the European North of the USSR (KB AS USSR 1983) pp. 25–38 (in Russian).

PANASENKO, G. D., and YAKOVLEV, V. M., About the nature of anomalous deformation of a transport tunnel in the mountain Yukspor. In Geophysical Investigations in the European North of the USSR (KB AS USSR, Apatity 1983) pp. 38–44 (in Russian).

REASENBURG, P. A., and SIMPSON, R. W. (1992), Response of regional seismicity to the Static Stress Changes Produced by Loma Prieta Earthquakes, Science 255, 1687–1689.

STEIN, R. S., KING, G. C. P., and LIN, J. (1992), Changes in Failure Stress on the Southern San Andreas Fault System Caused by the 1992 Magnitude = 7.4 Landers Earthquake, Science 258, 1328–1332.

YAKOVLEV, V. M., Recent crustal movement in the zone of the southern contact of the Khibiny Massif from data geometric levelling. In Geophysic and Geodynamic Investigations of the North-East of the Baltic Shield, (KB AS USSR, Apatitiy 1982) pp. 85–88 (in Russian).

(Received April 5, 1994, revised/accepted September 12, 1994)

PAGEOPH, Vol. 145, No. 1 (1995)

0033-4553/95/010039-19$1.50 + 0.20/0

The CQU Regional Seismic Network and Applications to Underground Mining in Central Queensland, Australia

Byron McKavanagh,[1] Bruce Boreham,[1] Kevin McCue,[1,2] Gary Gibson,[3] Jennifer Hafner,[1] and George Klenowski[4]

Abstract—The Central Queensland University (CQU) Regional Seismic Network is made up of an array of six short-period seismometer and two strong motion accelerometer stations. The array has an aperture of about 50 km. CQU is able to resolve epicentral co-ordinates to about ±2 kilometres, with a sample rate of 100 per second, and an absolute time accuracy of 100 milliseconds. This resolution is achieved by using triaxial seismometers which allow better secondary phase identification of shear, converted and depth phases.

The network covers an area of above average seismic risk in continental Australia. The area has been affected in 1918 by one of the largest earthquakes ever recorded along the eastern seaboard of Australia. The network also monitors a large number of blasts carried out by the coal mines and hard rock quarries in the region, and these are being used in a long-term study to determine the structure of the Crust and Upper Mantle in Central Queensland.

Techniques for monitoring rockbursts and longwall caving in mines are similar to those used for monitoring local earthquakes. CQU has successfully used a single triaxial seismometer to record seismic events produced by the strata failure and roof falls of a longwall coal mine. The case history presented shows that the initial fall under massive roof conditions appears to be predicted by a simple trend plot of microseismic event magnitude and longwall production rate. Extension of this technique to a closely spaced array of both surface and in-seam triaxial seismometers is required for a more detailed appraisal to be undertaken. Adequate resolution of event location requires a higher sample rate (up to 1000 Hz) and more accurate timing (about 1 millisecond) than for earthquake monitoring. An appropriate stratigraphic model is also required, as is the case for earthquake location.

Key words: Seismic network, mining-induced seismicity.

Introduction

This paper discusses the CQU Regional Seismic Network, the historical seismicity, and mining-induced seismicity of Central Queensland. The relatively brief

---

[1] Department of Applied Physics, Faculty of Applied Science, Central Queensland University, Rockhampton, Queensland 4702, Australia.

[2] Australian Seismological Centre, Australian Geological Survey Organisation, GPO Box 378, Canberra, ACT 2601, Australia.

[3] Seismology Research Centre, Royal Melbourne Institute of Technology, Bundoora, Victoria 3083, Australia.

[4] Capricorn Coal Development, Mail Bag, Middlemount, Queensland, 4714, Australia.

written history of this region shows that it is subject to damaging intraplate earthquakes. The Bundaberg 1918 earthquake, of M_L 6.3, has passed out of the oral history of the region, and an important part of our project is to ensure that the historical seismicity is taken into account when earthquake risk maps and building codes are under consideration. These events appear to occur due to the interaction of a relatively high horizontal tectonic stress, oriented north-northeast, and the northwest trending regional structural lineaments. Thrust faulting, or reactivation of suitably oriented normal faults as reverse faults is favoured in the current disposition of the stress field. This stress field is also perturbed by mining excavations (and large dams), and both blasting and mine-induced seismicity are studied by the Network.

The Central Queensland Regional Seismic Network

The network was established in May 1990 by the Central Queensland University (CQU), in collaboration with the Australian Geological Survey Organisation, the

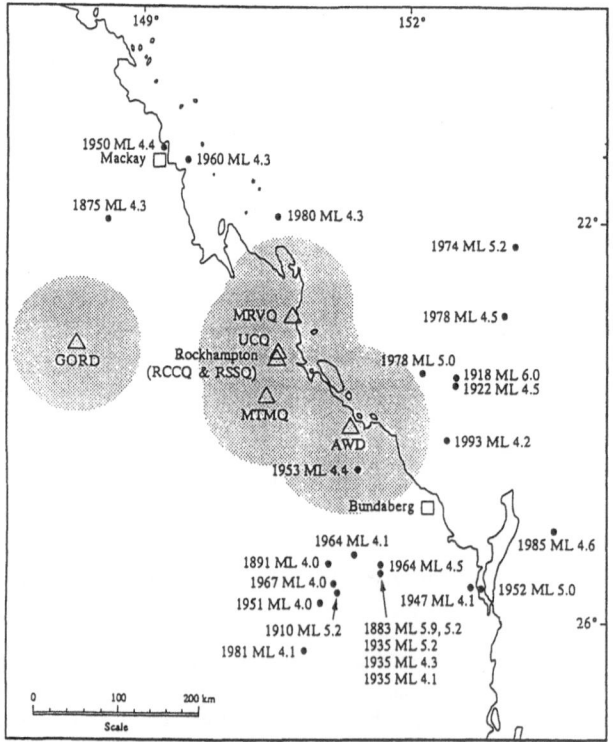

Figure 1

Seismograph sites, Network detection capability for magnitude M_L 2 and greater and Central Queensland historic earthquakes, magnitudes greater than M_L 4.

Table 1

Central Queensland seismograph stations

Name	Code	Location Lat.°S	Location Long°E	Elevation metres	Foundation	Instrument	Start Date
Kent Hill	UCQ3	23.3210	150.53383	120	Mudstone	MEQ-800	20.05.90
Maryvale	MRVQ	22.95484	150.67509	75	Granite	KELUNJI/S6000	09.04.92
Awoonga	AWD	24.0780	151.3157	110	Mudstone	MEQ-800	22.09.81
Fletcher Ck	MTMQ	23.76265	150.39014	170	Welded Tuff	KELUNJI/S6000	27.08.90
German Ck	GCM2	22.98	148.55	136	Permian Sediment	KELUNJI/S6000	01.11.90
City Council	RCCQ	23.368	150.513	20	Deep Soil	KELUNJI/ACCEL	17.06.93
Canning St.	RSSQ	23.387	150.50	25	Mudstone	KELUNJI/ACCEL	20.10.93

Queensland Department of Resource Industries, the Seismology Research Centre of RMIT and the University of Queensland. Essentially this was to provide detailed coverage of a populated area with significant historical seismicity. The early stages of Network establishment are detailed by COOPER *et al.* (1992). Network details are given in Table 1, while Figure 1 illustrates the locations of seismographs and the detection capability of the Network. The response curves of the network are shown in Figure 2. All but one of the digital seismograph (Kelunji) sites have modems connected to telephone lines, while one seismometer (UCQ3) is currently telemetered to an analog recorder on open display at the Applied Science office at CQU. Despite the obvious advantages of digital seismographs for data acquisition, processing and storage, a centrally located analog recorder is still useful for rapid assessment of local events. Once a preliminary epicentre and magnitude has been determined, either from S-P times from any two seismographs, or using the back azimuth calculation from a single Kelunji site (HAFNER *et al.*, 1994), felt reports can be sought via phone calls to nearby population centres. All network seismographs are checked, and collaborating networks are contacted to collate additional data and determine an accurate hypocentre using the program EQLOCL. A press release is issued for public information, and to request additional felt reports. If the epicentre is accessible, portable seismographs are re-located to record any aftershocks; such events are extremely valuable in improving the location of the main shock.

A simple 3-layer crustal model, SEQ2A, developed by Cuthbertson (1990), is presently in use. This is given as Table 2. A local earthquake catalogue is maintained at CQU, using the program ECATL to establish if any correlation exists between the current seismicity and tectonic structures, and to determine the regional stress field by constructing composite mechanism solutions.

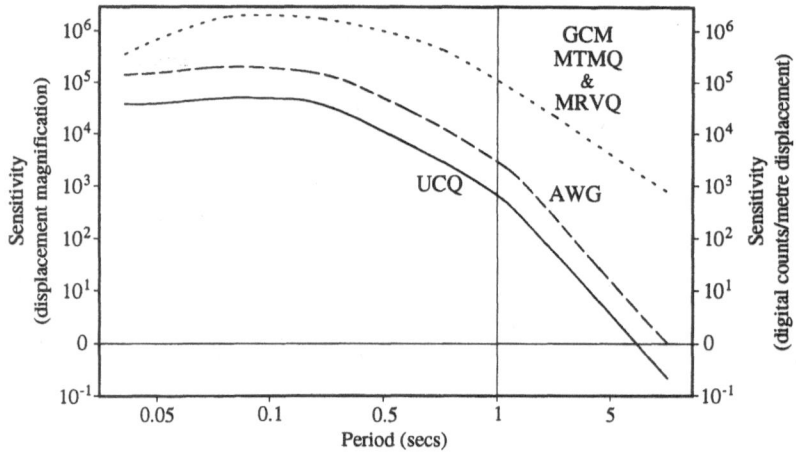

Figure 2
Central Queensland Regional Seismic Network seismograph response curves.

Table 2

Crustal model used for the CQU Regional Seismic Network

Depth (km)	Vp (km/sec)	Vs (km/sec)	Density (Mg/m³)	Q_P	Q_S
0 to 10	5.55	3.3	2.50	100	50
10 to 30	6.67	3.8	2.50	200	100
below 30	7.95	4.60	2.50	500	250

Seismicity of Central Queensland

The Central Queensland area has been shaken by at least fifteen earthquakes greater than magnitude M_L 4 in the last 110 years, as outlined in Table 3, and shown in Figure 1. Two of them were potentially very destructive with magnitudes of about 6; the 1883 Gayndah, and 1918 Bundaberg earthquakes (BRYAN and WATERHOUSE, 1938; EVERINGHAM *et al.*, 1982). Several of these earthquakes spawned large aftershocks. The 1883 Gayndah was followed by a M_L 5 aftershock. Within two hours of the Bundaberg 1918 event six aftershocks occurred with magnitudes in the range M_L 5.1–5.6, and aftershocks were felt in and around Rockhampton up to two months later. An isoseismal for the Bundaberg 1918 event

Table 3

Central Queensland Earthquakes M_L 4.0 or Greater

Date	Universal Time	Location Lat.°S	Location Long°E	Place	Magnitude M_L
28.08.1883	16:55	−25.5	151.7	Gayndah	5.9
24.11.1910	23:00	−25.7	151.2	Mundubbera	5.2
06.06.1918	18:15	−23.5	152.5	Bundaberg	6.3
07.03.1922	16:54	−23.5	152.5	Bundaberg	4.5
12.04.1935	01:32	−25.5	151.7	Gayndah	5.2
11.06.1947	10:03	−25.5	152.7	Maryborough	4.1
24.06.1952	01:44	−25.5	152.8	Maryborough	5.0
03.12.1953	15:42	−24.5	151.4	Manypeaks	4.4
19.10.1960	11:37	−21.2	149.5	Mackay	4.3
03.03.1964	06:13	−25.4	151.7	Gayndah	4.5
24.12.1974	02:25	−22.1	153.2	Coral Sea	5.2
28.11.1978	17:33	−23.36	152.43	Off Heron Island	5.0
08.02.1980	04:42	−21.8	150.5	N'Thumberland Is.	4.3
08.02.1985	08:23	−25.1	153.6	Off Indian Head	4.6
25.11.1993	04:07	−24.1	152.4	Lady Elliot Island	4.2

Data from Atlas of Isoseismal Maps of Australian Earthquakes BMR Bulletins #214, #222 and Queensland Department of Resource Industries Earthquakes Database.

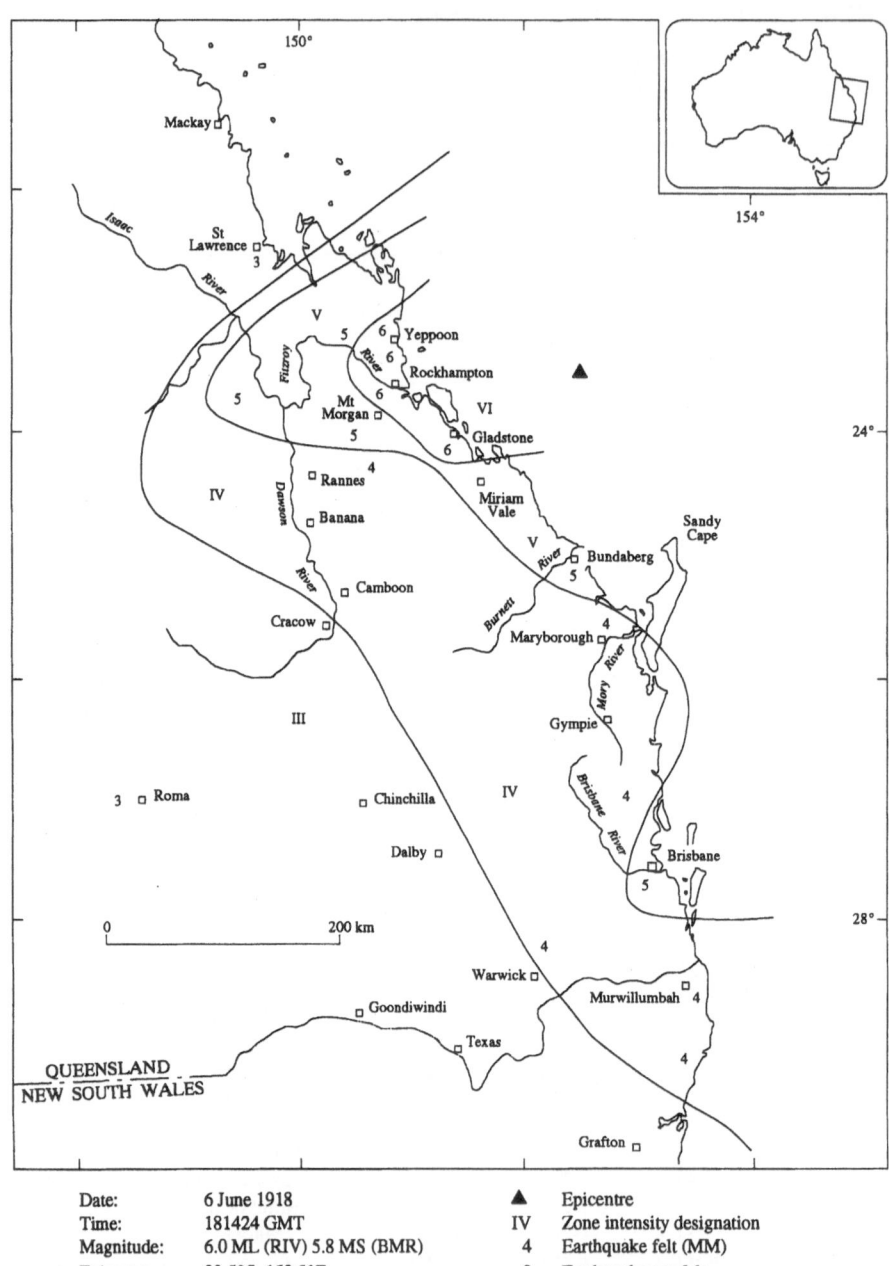

Date:	6 June 1918	▲	Epicentre
Time:	181424 GMT	IV	Zone intensity designation
Magnitude:	6.0 ML (RIV) 5.8 MS (BMR)	4	Earthquake felt (MM)
Epicentre:	23.5°S 152.5°E	0	Earthquake not felt

Figure 3
6 June 1918 major shock isoseismal map.

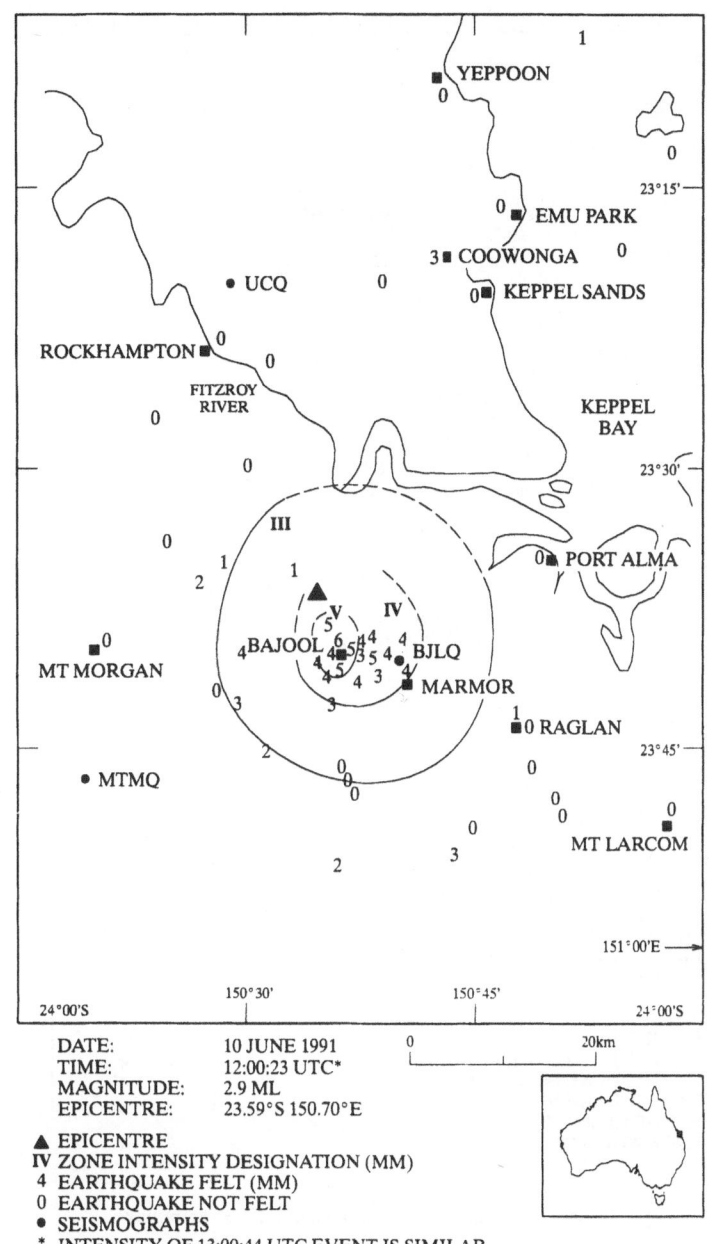

Figure 4
Isoseismal of the Bajool 1991 earthquake.

is shown in Figure 3, and from the radius of perceptibility, the magnitude has been revised to 6.3.

From the distribution of reported intensity and damage (COOPER *et al.*, 1992), it is now clear that both Rockhampton and Bundaberg experienced the damaging

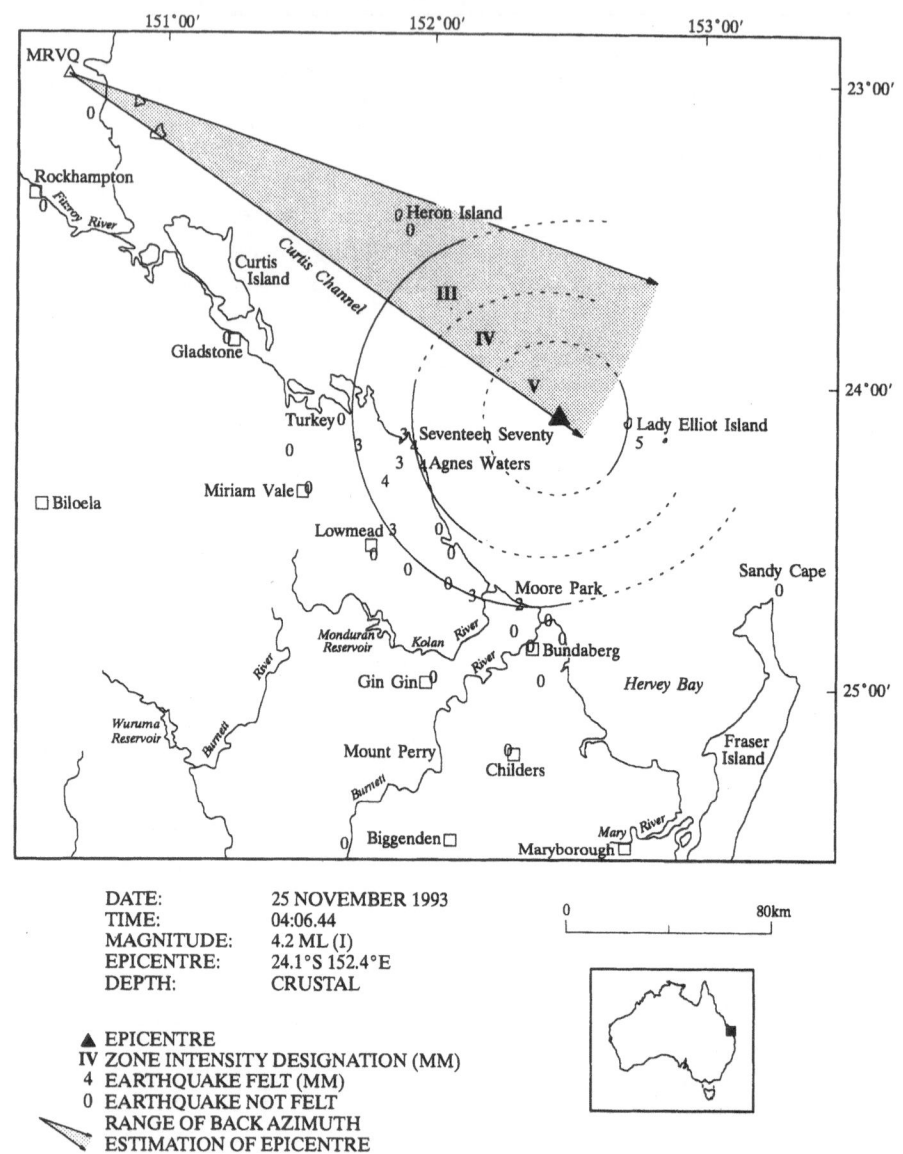

Figure 5
Isoseismal of the Lady Elliot Island 1993 earthquake.

effects of amplification of the seismic waves where buildings are founded on thick unconsolidated sediments. McCue *et al.* (1994) studied this effect for Rockhampton, and produced a zoning of seismic risk for the city.

Epicentres of earthquakes recorded by more accurate networks over the last 30 years indicate that the area is within the intersection region of two intraplate seismicity zones; a 500 km wide belt extending from Tasmania to Fraser Island, and from there to the tip of Cape York Peninsula. The Queensland Department of Resource Industries earthquake database, which incorporates information collected by the University of Queensland, indicates that the majority of earthquakes since 1977 with magnitude less than M_L 4 have occurred offshore of Central Queensland or to the south of an east-west line passing through Mount Morgan (CUTHBERT-SON, 1990). The area north of this line, including the coastal strip of Central Queensland, may be undergoing a period of quiescence, but it is more likely that due to lack of local seismographs, smaller earthquakes in this area have not previously been detected. Some small events have recently been recorded at station MRVQ, in apparent confirmation of this.

Two small earthquakes occurred near Bajool, on 10 June, 1991. These were recorded and felt widely around the epicentre (MCKAVANAGH et al., 1993a). The isoseismals of the larger event, M_L 2.8, are shown in Figure 4. Portable seismographs were installed in the epicentral region and they recorded several aftershocks.

An offshore earthquake of M_L 4.2 occurred on 25 November, 1993, near Lady Elliot Island. The back azimuth location (HAFNER et al., 1994) from station MRVQ proved useful in providing a quick epicentre determination, as shown in the isoseismal map of the event in Figure 5. Several smaller unlocated events have since occurred that could have come from the same epicentre, and these events highlight the need for extension of the network to the off-shore islands.

It is pertinent to note that there is no significant historical or instrumental record of mine-induced seismicity in the Central Queensland region. There are, however, numerous blasts in Central Queensland, which tend to dominate the record, as detailed below.

Central Queensland Blasting

The majority of seismic events recorded to date by the network has been from explosions, either quarry blasts or coal mine blasts. Coal mine locations are shown in Figure 6. Table 4 details blasting sites that have been recorded by the Network. We have learned to differentiate the blasts from earthquakes by their location, coda shape, time of day, and ultimately by confirmation of the blast times with the operators of the quarries and mines. The blasts always have large surface wave amplitudes relative to the P- and S-wave amplitudes, and sometimes surprisingly large S-wave amplitudes compared with those of the P wave. The coda shape of blast seismograms seems to be dependent upon several variables; quantity of explosives used, orientation of the blast in relation to the azimuth of the seismograph, and the blast delay interval.

Characteristic signatures of regional blasts have been identified to differentiate them from earthquakes, and a database of blast travel times established to improve

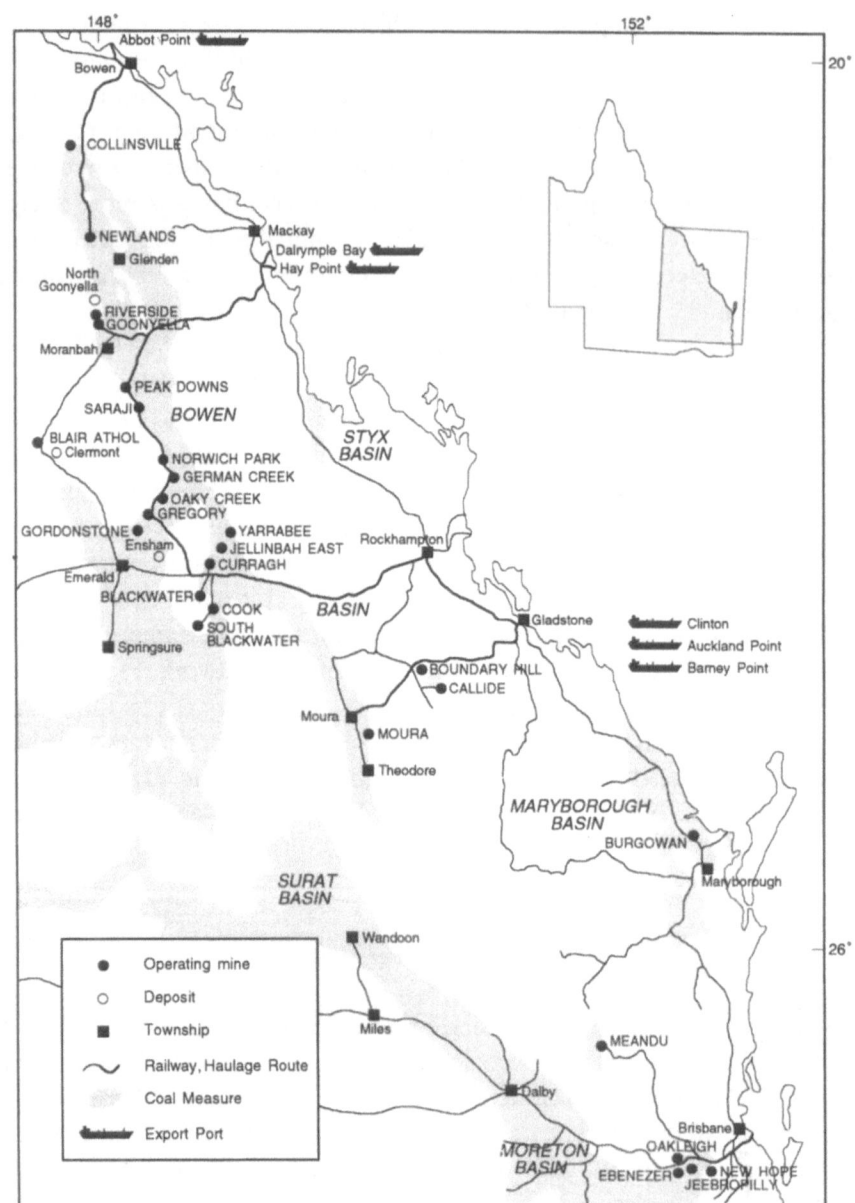

Figure 6
Location of Coal Mines and Infrastructure in Central Queensland.

the simple crustal model currently used for Central Queensland. Some blasts at Boundary Hill Mine have been accurately timed, and a repetition of this is intended at the other significant blasting sites, to determine accurate origin times for this database.

Table 4

Mines whose blasts are recorded on the Central Queensland Regional Seismic Network. Locations given for open-cut mines are approximate, as individual mines such as Peak Downs or Moura may be up to 20 km long. Note that we have also recorded blasts from South Blackwater, Newlands and Collinsville Mines

Name	Location Lat.°s	Long°E	Operators	Type
Nerimbera	−23.40	150.60	SELLARS QUARRY	Surface Quarry
Mt. Etna	−23.16	150.45	CENTRAL QLD CEMENT	Surface Quarry
Marmor	−23.68	150.72	WELLS LIME WORKS	Surface Quarry
Taragoola	−24.10	151.24	FROST ENTERPRISES	Surface Quarry
Callide	−24.32	150.63	CALLIDE COAL	Open Cut
Boundary Hill	−24.30	150.59	CALLIDE COAL	Open Cut
Moura	−24.61	150.09	BHPAC	Open Cut
Blackwater	−23.77	148.83	BHPAC, CURRAGH	Open Cut
Norwich Park	−22.70	148.42	BHPAC	Open Cut
Peak Downs	−22.23	148.17	BHPAC	Open Cut
Saraji	−22.39	148.27	BHPAC	Open Cut
German Creek	−22.98	148.55	CAPCOAL	Open Cut and Longwall
Oaky Creek	−22.85	148.50	MT ISA MINES	Open Cut and Longwall
Gregory	−23.90	148.17	BHPAC	Open Cut
Goonyella	−21.78	147.92	BHPAC	Open Cut

Applications to Underground Mining in Central Queensland

The future of the export coal industry in the Bowen Basin of Central Queensland is based increasingly on deeper reserves of coal. There are several approaches to the extraction of these reserves, ranging from longwall mining of deeper, gassy coal in thick seams, to the proposed 90 m highwall open cuts where overburden and seam conditions are favourable. In each of these mining methods there is a potential application of either standard seismic, or microseismic techniques to enhance safety and productivity. The principal advantage of such techniques bears re-iterating: because they operate on self-generated seismic signals, not only can they be measured remotely from the area of operation, but they are also sensitive indicators in space and time of the failure condition of the rock mass. STYLES et al. (1992a) reviewed applications of microseismic techniques to the mining industry globally. There is a wide variety of systems installed, varying from the highly developed and semi-automated Integrated Seismic System, installed in several deep South African gold mines, to the simplest possible system, a single detector used to warn of falls and bursts in the working area of Chinese mines.

In terms of the application to longwall monitoring, an interesting approach is described by KONECNY' (1992) in the Upper Silesian Basin of Czechoslovakia. Data from a local, mine-based network is combined with a regional seismic network to evaluate mine-induced seismicity in multi-seam extraction. Other papers describe permanent seismic networks applied to the coal mining industry in West Germany (WILL and RAKERS, 1992), France (BEN SLIMAN and REVALOR, 1992), Canada (HASEGAWA *et al.*, 1989), and Poland (see STYLES *et al.*, 1992a), while other coal mining countries have used microseismic monitoring more on a research basis, e.g., USA, Australia (see STYLES *et al.*, 1992b; and GIBOWICZ, 1984). There are more permanent installations worldwide in deep hard rock mines, as detailed in STYLES *et al.* (1992a).

WONG (1993) reviewed available focal mechanisms of mine-induced seismicity to evaluate the effects of tectonic stresses. He found that compressive tectonic stresses appear to play a major role in the generation of mine-induced seismicity worldwide. Since the Bowen Basin in Central Queensland is subjected to a relatively high compressive horizontal stress reaching three times the vertical stress (ENEVER, 1990), it is expected that significant mine-induced seismicity will occur as longwall mines progress deeper. Presently, smaller mine-induced events (microseismic activity) are being monitored to study their relationship to mining. The term "microseismic" is used in this context as meaning a smaller, higher frequency event than a regional earthquake. Note that this is somewhat different than the use of the term in earthquake seismology, but it is common usage in mining engineering.

The concept of monitoring longwall caving characteristics by seismic methods from the surface was trialled by Central Queensland University at Capcoal's Southern Colliery, German Creek in late 1990 (McKAVANAGH *et al.*, 1993b). Figure 7 shows the location of the seismograph above the longwall panels in plan

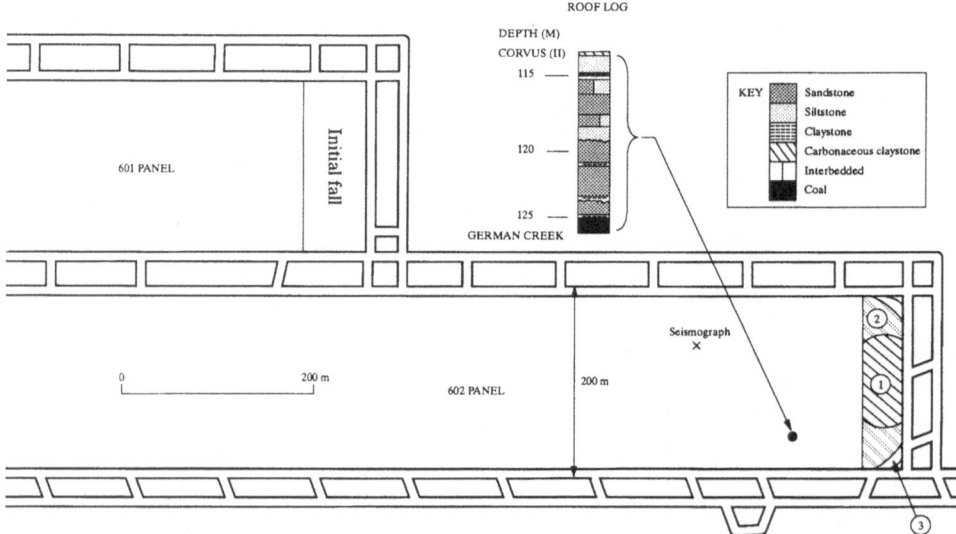

Figure 7
Panel layout in plan at Southern Colliery, German Creek showing initial falls in 601 and 602 Panels, and position of the surface seismograph.

view, and the extent of the initial falls in both panels. The depth of cover averages about 125 m in Panel 602. Previous experience established when Panel 601 commenced, the initial fall was extensive, owing to the massive nature of the immediate roof above the extracted seam, as shown in the roof lithological log in Figure 7. The surface mounted seismograph produced useful data, enabling seismic characterisation of several caving sequences as 602 longwall commenced.

The seismograph consisted of a Kelunji digital recorder and a Sprengnether S-6000 three-component short-period (2 Hertz) transducer. The transducer was installed on a pedestal of cement grout, backfilling a hole drilled 8 metres into competent Permian sediments. The recorder sampled at 200 times per second, using a short time averaging/long time averaging trigger algorithm to select seismic events to record, and to disregard background noises of a more continuous nature. A total of 315 events were recorded, of which 152 are considered to be longwall related seismic events. These events have been analysed and plotted, using the program "SeisMac" on the C.Q. Regional Seismic Network's McIntosh IIsi computer. An energy index has been calculated from the sum of the squares of the peak amplitudes

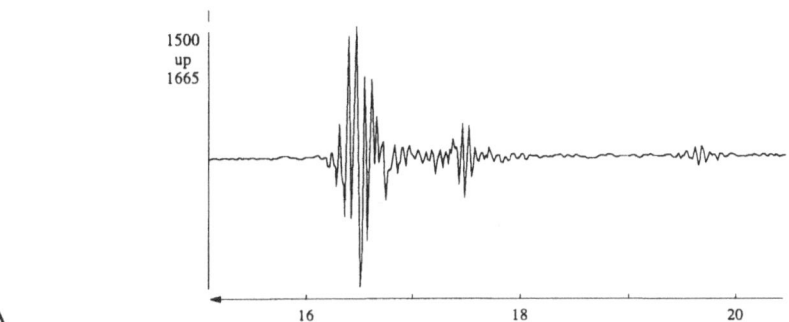

A

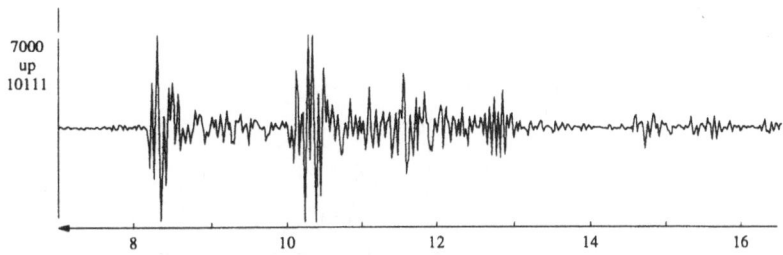

B

Figure 8

A and B: These waveforms are a plot of velocity amplitude against time in seconds, as recorded on the seismograph above 602 Panel, Southern Colliery, German Creek. The peak amplitude of each waveform is shown on the *y* axis. A is a typical type 1 (microseismic) event, while B is a complex waveform (type 2) associated with falls.

of the components of the waveform, but is expressed in arbitrary units as we have not allowed for the effects of attenuation or variations in distance between the source and the transducer. For the geometry of the installation, we estimate the distances will vary from a minimum of 100 m, to a maximum of about 200 m, and that this will cause errors in the energy index ranging from $+100\%$ typical to $+400\%$ worst case between recorded events.

The seismic events were classified into four types, defined principally by waveform analysis. These are:

1. *Strata failure (microseismic events)*: i.e., the internal failure of roof layers, without a subsequent fall. An example is shown in Figure 8a. As can be seen, these are characterised by an impulsive nature, i.e., the waveform manifests a distinct and abrupt commencement. The events are also short in duration, 2 seconds being typical, and there tends to be a decrease in frequency content with time during the waveform. Figure 9 is a spectral analysis of a type 1 waveform, from which a

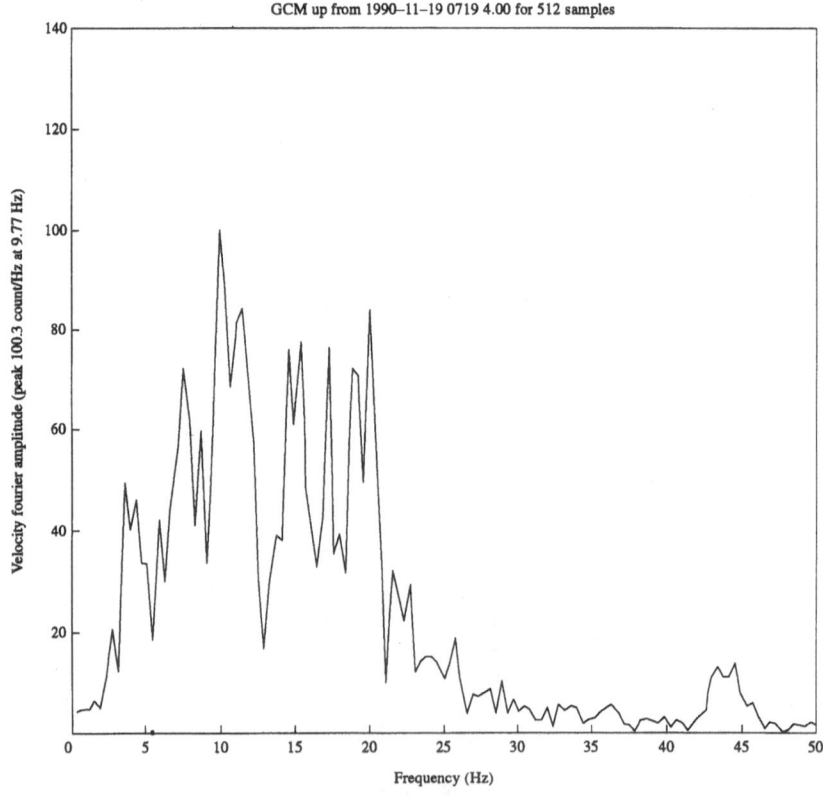

GCM up from 1990–11–19 0719 4.00 for 512 samples

Figure 9

Spectral analysis of the type 1 waveform shown in Figure 8A. From the "corner frequency" of 23 Hz, a "source radius" of 35 m can be calculated. Seismic moment, and stress drop can also be calculated from this type of analysis.

"source radius" of 35 m has been calculated, using the source models of BRUNE (1970) and MADARAIGA (1976).

2. *Falls and associated seismic noises*: These include the strata breakage, frictionally generated noises as the blocks begin to move, and possible impact noises as the roof blocks fall onto the floor or other roof blocks. These were recognised by time of occurrence and subsequent correlation to observed and inferred fall sequences. These were very often multiple, closely-spaced events, as shown in Figure 8b.

3. *Production related seismic noises*, such as accidental impacts, shearer cutting roof, coal bumps, etc. These are difficult to identify without close communications with personnel on the face, but since they are generally much smaller than roof noise, they should not extensively contaminate the data.

4. *Switching spikes* in electrical power lines. These are distinctive and easily removed.

There were seven sequences of microseismic events between 16 November and 20 November, 1990 that are considered to be of significance in the monitoring of 602 Panel. These are listed in Table 5 together with our interpreted fall times. (Fall times were recorded where possible by mining personnel who were present, but the accuracy was limited ± 5 minutes. Mining personnel were not continuously present in the Panel from 23:00, 16 November to 07:00, 19 November because the major fall was expected. Fall times were recorded by staff in adjacent Panels, and confirmed by brief inspections. Note that we were not initially advised of these times, and had to prove to the satisfaction of the Senior Mining Engineer that we had recorded the fall times accurately). These sequences are characterised by a progressive increase in amplitude and a shortening of time between type 1 microseismic events. This was followed by a series of sharp multiple events (type 2) occurring over a short period of time, usually less than one minute. The largest event recorded in this period is shown in Figure 8b. This shows a type 2 event, immediately preceded by a type 1 event. Such complex waveforms appear to be

Table 5

Observed sequences of microseismic events and roof falls, 602 Panel, Southern Colliery, German Creek. All times are local time, (Australian Eastern Standard Time)

Sequence	Date	Times	Number of Events	Fall Time
1	16/11/90	12:54–14:52	8	16:11
2	17/11/90	05:24–06:21	7	6:19 and 7:15
3	17/11/90	14:40	1	17:01 and 17:32
4	17/11/90	18:56–19:29	10	19:28 and 19:36
5	17/11/90	19:50–20:14	6	23:36
6	18/11/90	01:05–02:23	4	07:24 and 08:08
7	19/11/90	17:08–18:23	8	18:21

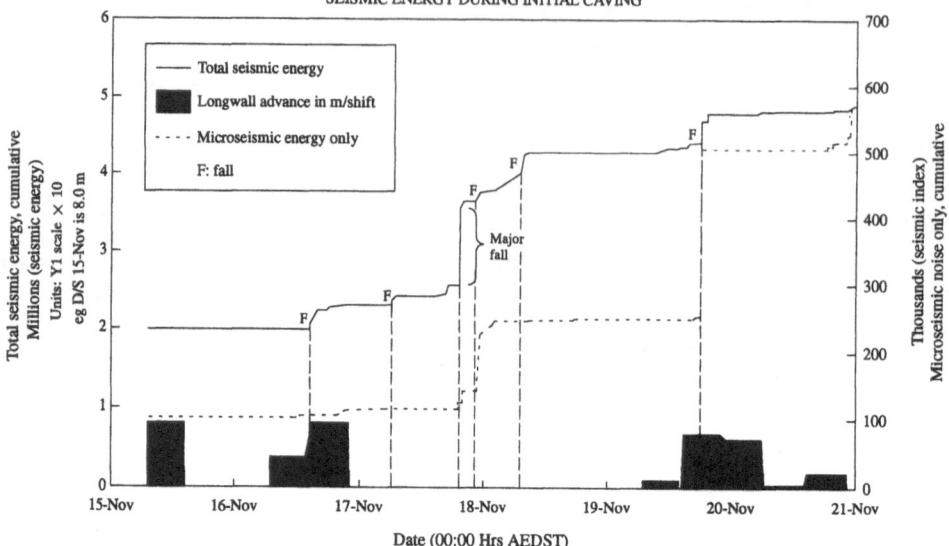

Figure 10
Total seismic energy, microseismic energy and coal production shifts during the monitoring period for
Panel 602, Southern Colliery, German Creek.

typical of falls. The events are summarised in Figure 10, which shows a cumulative
total of all seismsic noise, microseismic noise and longwall production per shift.

It is apparent from the trend plot of the microseismic noise, labelled strata noise
in Figure 11, that these nosies demonstrate a characteristic exponential increase in
cumulative energy before the largest roof fall. Note the increase in microseismic
noise in the 30 minutes preceding the falls at 19:36 hrs. It should also be noted that
the microseismic noise increase at 20:00 hrs was not followed by falls, so that this
simplistic approach will obviously require further refinement before a prediction
technique is developed. With this in mind, we are studying the digitally recorded
waveforms in greater detail, to determine if there are characteristic changes in event
magnitude, frequency content, or in "signature" with time preceding the falls. We
are also upgrading to a network of triaxial seismometers to locate the events, so as
to reduce the large errors in the seismic energy index.

From this and subsequent monitoring at Southern Colliery, it has been concluded that the digital seismograph has proven to be reliable, easily operated and
installed, and of excellent sensitivity for this type of monitoring. Falls during initial
caving were readily recognisable from the recorded waveforms, and could be
correlated to the extensive observations of fall times and character, despite the
subjective nature of some of this data. The sequences of microseismic events
observed to precede the falls would prove useful as an objective measure of the
amount of roof noise routinely heard underground. This form of monitoring

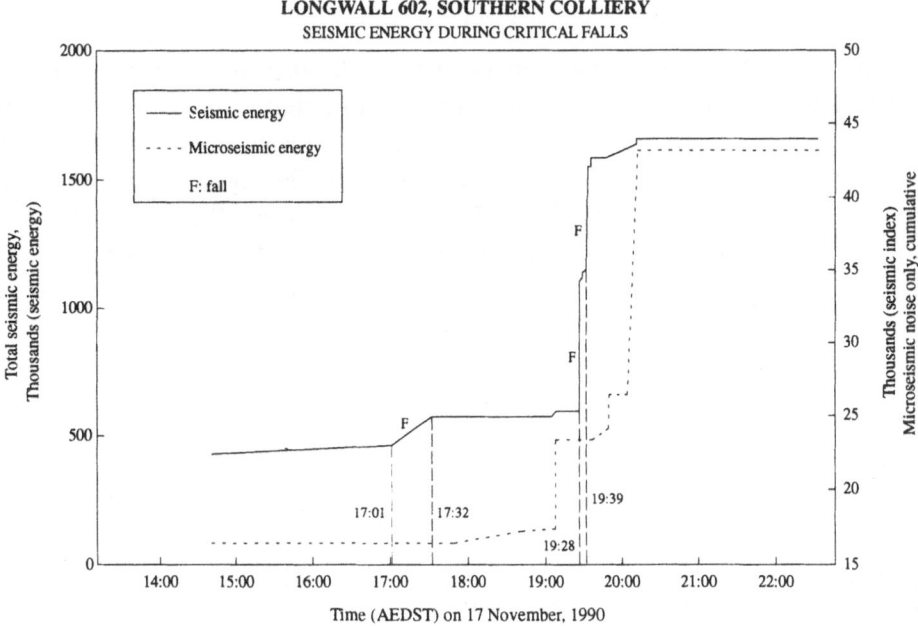

Figure 11
Total seismic energy, microseismic energy during the critical stage of mining 602 Panel, Southern Colliery, German Creek.

promises to provide warning of longwall caving problems, such as initial caving under massive sandstone roof.

Conclusion

Now that the Central Queensland Network has been established and is operating routinely, it is our intention to concentrate on the understanding and modelling of problems in local seismicity which have national and international significance. We will concentrate on increasing our understanding of the causes and mechanisms of intraplate earthquakes, and we will also apply the network to mining problems in an effort to improve mine safety. Seismic characterisation of longwall mining appears to have potential in providing useful information on the behaviour of the roof strata over the area of extraction, for which there are few other available techniques. We expect that mine-induced tectonic events will also occur, and it would be interesting to study the relationship of these with the microseismic activity induced closer to the mine openings. Source mechanisms of such events will provide information on the tectonic stresses around the mines.

Acknowledgements

The authors thank the Shell Company of Australia as managers of the Capricorn Coal Development Joint Venture for permission to publish this paper. Substantial funding for the German Creek seismograph has been provided by a research grant from Central Queensland University. The Australian Geological Survey Organisation, the Queensland Department of Resource Industries, Seismology Research Centre of RMIT, and the University of Queensland are all thankfully acknowledged for their substantial contributions to, and collaboration with, the CQ Regional Seismic Network.

REFERENCES

BEN SLIMAN, K., and REVALOR, R., *Some results on the accuracy of the location of mining-induced seismic events: Experience of French coal mines.* In *Induced Seismicity* (ed. Knoll, P.) (Balkema, Rotterdam 1992) pp. 3–20.

BOLLINGER, G. A., *Microearthquake activity associated with underground coal-mining in Buchanan County, Virginia, U.S.A.* In *Seismicity in Mines* (ed. Gibowicz, S. J.) (Birkhäuser Verlag, Berlin 1989) p. 407.

BRUNE, J. N. (1970), *Tectonic Stress and the Spectra of Seismic Shear Waves from Earthquakes,* J. Geophys. Res. *75*, 4997–5009.

BRYAN, W. H., and WHITEHOUSE, F. W. (1938), *The Gayndah Earthquake of 1935,* Proceedings of the Royal Society of Queensland *49*, 106–119.

COOPER, W. V., MCKAVANAGH, B. M., BOREHAM, B. W., MCCUE, K., CUTHBERTSON, R. J., and GIBSON, G. (1992), *The Regional Seismographic Network and Seismicity of Central Queensland,* BMR J. Australian Geology and Geophys. *13*, 107–111.

CUTHBERTSON, R. J. (1990), *The Seismo-tectonics of Southeast Queensland,* Bureau of Mineral Resources Bulletin, Australia, *232*, 67–81.

CUTHBERTSON, R. J. (1989), *Geological Implications of Eastern Queensland Seismicity and Focal Mechanisms of Earthquakes in Southeast Queensland,* Bureau of Mineral Resources Bulletin, Australia, Record 1989, 6.

ENEVER, J. R. (1990), *"In situ stress measurements in the Bowen Basin and their implications for coal mining and methane extraction."* In *Proceedings, Bowen Basin Symposium 1990* (ed. Beeston, J.W.) Geological Society of Australia (Old Divison).

EVERINGHAM, I. B., MCEWAN, A. J., and DENHAM, D. (1982), *Atlas of Isoseismal Maps of Australian Earthquakes,* Bureau of Mineral Resources Bulletin, Australia 214.

GIBOWICZ, S. J. (1984), *The mechanism of large mining tremors in Poland.* In *Rockbursts and Seismicity in Mines* (eds Gray, N. C. and Wainright, E. H.), South Africa Institute Min. Metall., Johannesburg, Symp. Sev. *6*, 17–28.

HAFNER, J. K., BOREHAM, B. W., MCKAVANAGH, B. M., and JEPSEN, D. (1994), *Application of a principal component technique with single station three-component digital data to back azimuth estimation of some aftershocks of the Tennant Creek earthquakes of 1988.* In Abstracts of IASPEI 27th General Assembly, January 10–21 1994, Wellington, New Zealand, S10.18.

HASEGAWA, H. S., WETMILLER, R. J., and GENDZWILL, D. J., *Induced seismicity in mines in Canada—an overview.* In *Seismicity in Mines* (ed. Gibowicz, S. J.) (Birkhäuser Verlag, Berlin 1989) p. 423.

KONECNY', P., *Mining-induced seismicity (rock bursts) in the Ostrava-Karvina' Coal Basin, Czechoslovakia.* In *Induced Seismicity* (ed. Knoll, P.) (Balkema, Rotterdam 1992) pp. 107–130.

MADARAIGA, R. (1976), *Dynamics of an Expanding Circular Fault,* Bull. Seismol. Soc. Am. *66*, 639–666.

McCue, K. F., Boreham, B. W., McKavanagh, B. M., and Bugden, C. (1994), *Amplification of earthquake ground motion by alluvial sediments in Rockhampton, Central Queensland, Australia*. In Abstracts of IASPEI 27th General Assembly, January 10–21 1994, Wellington, New Zealand, W8.15.

McKavanagh, B. M., Boreham, B. W., Cuthbertson, R. J., McCue, K. F., and Cooper, W. V. (1993a), *The Bajool Earthquake Sequence of 1991, and Implications for the Seismicity of Central Queensland*, Australian J. Earth Sci. 40, 455–460.

McKavanagh, B. M., Boreham, B. W., Cuthbertson, R. J., Gibson, G, Klenowski, G., and McCue, K. F. (1993b), *Underground mining applications of the UCQ regional seismic network*. In *Geotechnical Instrumentation and Monitoring in Open Pit and Underground Mining* (ed. Szwedzicki) (Balkema, Rotterdam, 1993).

Rynn, J. M. W. *et al.* (1987), *Atlas of Isoseismal Maps of Australian Earthquakes*, Bureau of Mineral Resources, Australia, Bulletin 222.

Styles, P., Mallett, C. W., and Toon, S. M. (1992a), *The Application of Microseismic Techniques, and the Potential for Application to Australian Mines*, CSIRO Division of Geomechanics CMTE Report, October.

Styles, P., Bishop, I, Toon, S., and Trueman, R. (1992b), *Surface and Borehole Microseismic Monitoring of Longwall Faces; Their Potential for Three-dimensional Fracture Imaging and the Geomechanical Implications*, 11th International Conference on Ground Control in Mining, Australian Institute of Mining and Metallurgy, Wollongong, June 1992.

Will, M., and Rakers, E., *Induced seismoacoustic events in burst-prone areas of West German coal mines*, In *Induced Seismsicity* (ed. Knoll, P.) (Balkema, Rotterdam 1992) pp. 185–212.

Wong, I. G., *Tectonic stresses in mine seismicity: Are they significant?* In *Rockbursts and Seismicity in Mines* (ed. Young, P.) (Balkema, Rotterdam 1993) pp. 273–278.

(Received May 24, 1994, revised/accepted February 6, 1995)

PAGEOPH, Vol. 145, No. 1 (1995)

0033–4553/95/010059–10$1.50 + 0.20/0

Research on Earthquakes Induced by Water Injection in China

Zhao Genmo,[1] Chen Huaran,[1] Ma Shuqin[1] and Zhang Deyuan[2]

Abstract — In China, the earthquakes induced by water injection have occurred in four oil fields including the Renqiu oil field, and in two mines. Production of oil from the Renqiu oil field began in 1975 and the injection of water into the oil field commenced in July 1976. The induced earthquakes have been occurring in the area for the past 17 years, since December 1976. The controlled experiments of water injection showed the cause and effect relation between water injection and earthquakes. Source parameters such as source dimension, seismic moment and stress drop of a large number of the induced earthquakes, and Q factor for the area have been determined. The results indicate that the stress drop varies from 0.2 to 3.0 bar and the Q factor has an average value of 75.0. The low-stress drop and low Q factor values imply that the earthquakes are caused by the brittle fracture of weak rocks under low ambient stresses, due to a decrease in their strength because of the injection of water. The induced earthquakes are unevenly distributed in the oil field. The northern part of the oil field, where the reservoir rocks are characterized by low porosity and low permeability, exhibits high seismic activity with the largest earthquake registering a magnitude of 4.5 and about 68% of the total number of induced earthquakes in this part. Whereas, the southern part of the oil field with higher porosity and higher permeability is characterized by low seismic activity with the largest earthquake registering a magnitude of 2.5 and only 4% of the total number of earthquakes which occurred in this part. These features of the focal region suggest that larger earthquakes may not occur in the Renqiu oil field area.

Key words: Earthquakes, water injection, oil field.

Introduction

The earthquakes induced by water injection have been reported by Evans (1967), Healy *et al.* (1970) and Rayleigh *et al.* (1976). These studies have attracted great interest amongst Chinese seismologists. The earthquakes induced by water injection were detected from an oil field in east China. About 85.4% of the total output of oil comes from such oil fields. The techniques of water injection are being applied for stimulating oil fields in China. Until now, the earthquakes which induced water by injection have occurred in four oil fields with a maximum intensity reaching MM VI and in two rock salt mines with the maximum intensity reaching MM VII. These earthquakes have caused serious damage to the buildings

---

[1] Seismological Bureau of Tianjin 300201, PR China.
[2] Oil Field Bureau of North China, Tianjin 300280, PR China.

in the nearby areas. Some earthquakes are also found to be induced due to underground disposal of nuclear waste from the first nuclear power plant in China.

The Relation Between the Earthquakes and Water Injection in the Oil Fields in North China

Renqiu oil field in the central part of the China plain, situated at a distance of 150 km south of Beijing, is a typical case of induced earthquakes, in an oil field in North China. This oil field is about a 20-km long and 6-km wide strip. The upper layer of the crust in the region consists of about 8-km thick mudstones, sandstones and limestones of Cenozoic, the sub-Paleozoic and the sub-Sinian eras, while the lower part is about a 20- to 30-km thick base of metamorphic rocks, granites migmatites and basalts. The oil reservoir is at a depth of about 3000 m to 4000 m, the structure of which is a series of NNE trending anticlines and synclines of thick carbonate rocks of the sub-Sinian and the sub-Paleozoic Eras, and is cut by a fault into broken lump hills of half anticlines or monoclines. Eighty-five faults have been found in the oil field, and on the western boundary of the oil field there is a 21-km normal fault (Ren-Xi fault). The Beijing-Tangshan seismic zone is at a distance of 150–180 km to the north of the oil field, the Xingtai seismic zone is 120 km to the south and the Hejian-Litan seismic zone about 20 km to the east. After the Hejian earthquake of magnitude 6.3 in March, 1967 four small earthquakes occurred in Renqiu area. From 1969 through 1976 no earthquake was detected. The Renqiu oil field belongs to an aseismic zone in the north China plain (Figs. 1 and 2).

In the Renqiu oil field, oil extraction began in 1975 and water injection in December 1976. The amount of water injected was about 1 million cubic meters per month in 1977, up to 1.5 million cubic meters in 1980 and then down to 1.1–1.0 million cubic meters from 1981 to 1983. The depth of water injection was about 3000–4000 m. The recording of earthquakes began in March 1977. With the seismic stations located in Wanxian and Wenan counties belonging to the Hebei province, the capability of this network to monitor earthquakes from this oil field was very low. Hence, in 1980 with the cooperation of the Geophysical Research Institute of the State Seismological Bureau, the network for monitoring the seismicity of the oil field was installed. Three-component short-period seismic detectors were installed in five wells of 500–1000 m deep. The seismic signals were transmitted to the stations through cables. In 1988 this system was formally put into operation for the transmission of signals to the Beijing network and the digital recording facility was added in 1990. From March 1977 to 1987, 300 earthquakes were recorded, nine of them attained a magnitude over 3.0 and the largest of them reached a magnitude of 4.5. The focii of these earthquakes are shallower, between 3–5 km deep and the earthquakes could be felt, with a maximum intensity of MMV–VI. Generally, the shocks of magnitude over 2.0 can be felt noticeably with buildings shaking, and

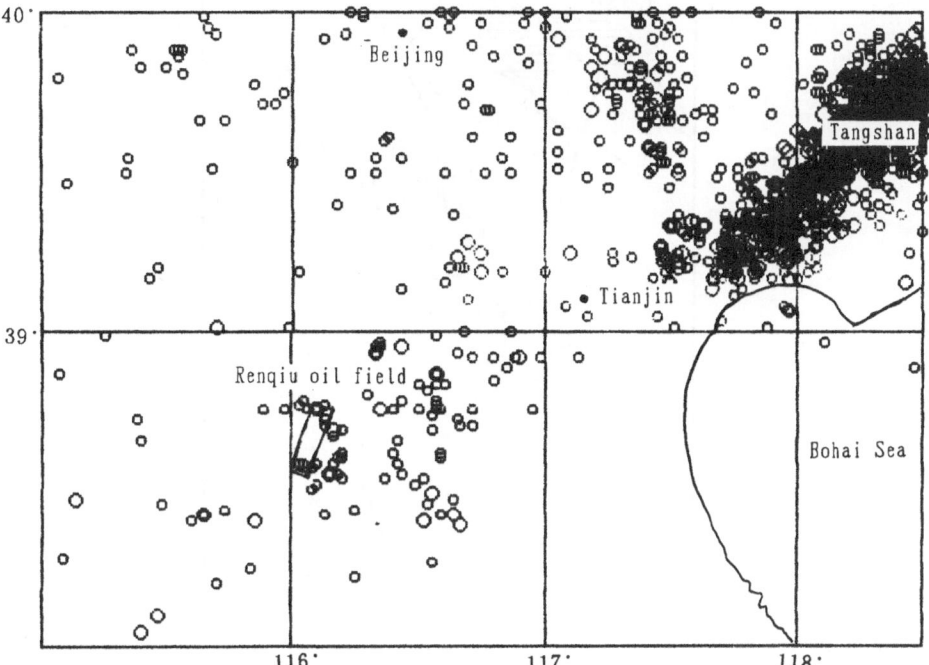

Figure 1
The location of the Renqiu oil field and the neighboring seismic zone.

doors and windows rattling. These induced earthquakes are distributed around the water injection wells. The total amount of injected water is correlated with the total frequency of earthquakes (Figs. 3 and 4).

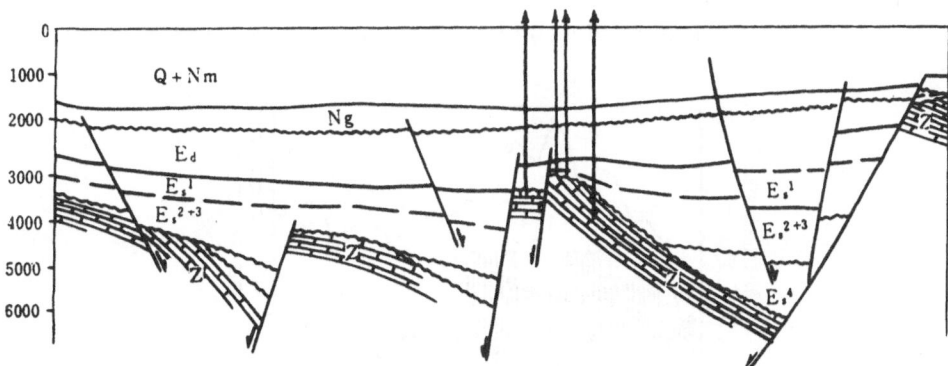

Figure 2
The geological cross section of the Renqiu oil field.

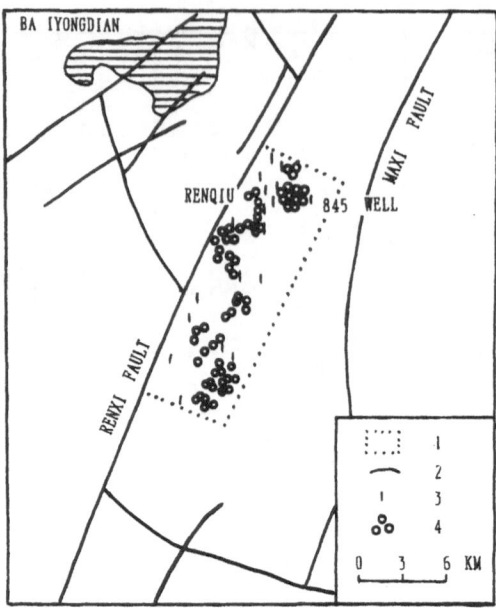

Figure 3
Location of faults, injection wells and epicenter of earthquakes in the Renqiu oil field. 1. Oil field, 2. faults, 3. injection wells and 4. epicenter.

In 1988, through a 4200-m deep new well (No. 845), water was injected into the oil reservoir of limestone rocks of the Ordovician Era in the northern part of the oil field. Thereafter a group of concentrated earthquakes were induced, the activity being maximum for a single water injection well in this oil field. The injection began on July 9, 1986 at the rate of 300 cubic meters of water per day and the injection pressure was 33.8 MPa. Earthquakes were recorded from September 6 onwards and by December they had increased 56 times. To detect the cause of these earthquakes, injection of the water was stopped on December 12 and the seismic activity ceased immediately. Water was injected again from February 5, 1987 onwards at the rate

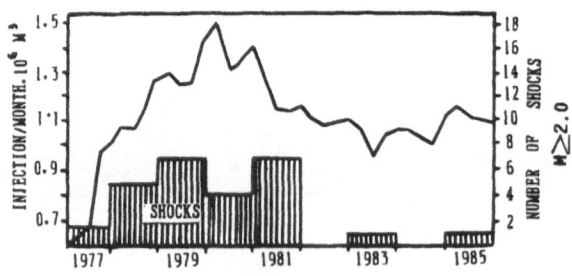

Figure 4
Frequency of earthquakes as a function of water injection per month in the Renqiu oil field.

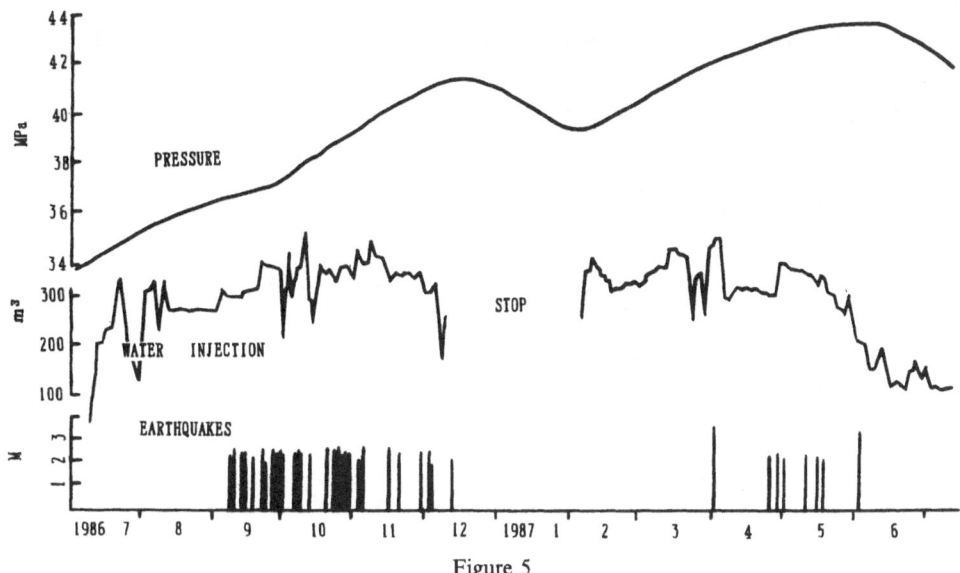

Figure 5
The relation between the water injection in well 845 and the swarm of earthquakes.

of 300 cubic meters per day and at an injection pressure of 38.5 MPa. On March 30, earthquakes were recorded when the injection pressure was 41.85 MPa. On June 2, the largest one reaching a magnitude of 4.5 occurred, when the injection pressure was 43.26 MPa. From June 2 the rate of water injection decreased to 150 cubic meters a day and the seismicity almost stopped. This demonstrates the cause and effect relation between the water injection into the well 845 and the swarm of earthquake activity (Fig. 5). This concentrated swarm of earthquakes, located near a building of the oil field at a distance of 1.5 km to the west of well 845, caused the collapse of the roof. The difference in the arrival time of P and S waves at well 104 was within 0.6–0.8 s and the focal depth was estimated to be about 3 km. The b value for these earthquakes is 1.2, close to the one for a swarm of reservoir-induced earthquakes or for a swarm of natural earthquakes.

The Focal Parameters and the Medium Characteristics in the Oil Field

Though the correlation between the earthquakes and water injection is very noticeable in the oil field, people were deeply worried about them. It is not known whether it is possible to induce still larger earthquakes due to continued injection of water, leading to a more serious disaster. This apprehension is justified for the following reasons:

Suddenly a discarded well in this oil field gushed 8 days earlier than the Tangshan M 7.8 earthquake on July 28, 1976, long before the water injection

Table 1

Focal parameters

No.	T (yr)	m	d	h	min	s	0_N	0_E	M_L	r (km)	t_{2a} (s)	U_{am} (μm)	a (km)	$M_0 \times 10^{14}$ (Nm)	$\Delta\sigma$ (bar)	Δu (cm)	Q_0	Q_c
1	1986	9	06	16	46	48.6	38.45	116.15	2.2	14.8								53
2	1986	9	07	01	07	47.9	38.45	116.15	2.7	19.6	0.22	0.05	0.25	0.73	0.2	0.01	98	81
3	1986	9	09	13	39	44.4	38.43	116.14	2.2	20.6	0.15	0.08	0.168	0.83	0.72	0.03	187	62
4	1986	9	13	05	01	21.8	38.42	116.19	2.2	22.6	0.12	0.12	0.13	1.05	2.30	0.06	251	89
5	1986	9	14	20	35	06.1	38.41	116.08	2.5	07.8								9
6	1986	9	17	00	40	01.9	38.44	116.17	2.0	18.6	0.17	0.06	0.19	0.63	0.39	0.05	143	64
7	1986	9	22	08	40	47.0	38.41	116.15	2.1	26.5	0.19	0.10	0.21	1.65	0.80	0.037	176	61
8	1986	9	25	04	48	36.1	38.41	116.12	2.5	28.4	0.19	0.2	0.21	3.56	1.73	0.079	189	69
9	1986	9	26	09	17	21.7	38.42	116.12	1.9	27.5	0.18	0.1	0.20	1.64	0.89	0.039	196	81
10	1986	9	27	23	48	54.2	38.44	116.14	2.4	20.6	0.16	0.17	0.18	1.87	1.37	0.054	171	75
11	1986	9	29	10	32	29.3	38.45	116.14	2.6	20.6	0.19	0.22	0.21	2.83	1.38	0.063	187	67
12	1986	10	05	08	41	29.1	38.37	116.10	2.3	34.4	0.14	0.06	0.16	0.98	1.07	0.037	344	51
13	1986	10	08	09	34	44.2	38.44	116.15	2.3	20.6	0.16	0.1	0.18	1.10	0.81	0.032	187	59
14	1986	10	09	03	47	56.2	38.44	116.14	2.2	20.6	0.17	0.14	0.19	1.63	1.02	0.042	158	56
15	1986	10	12	11	18	11.4	38.43	116.13	2.6	24.5	0.19	0.20	0.21	3.07	1.49	0.068	163	61
16	1986	10	19	12	27	42.1	38.40	116.12	2.9	13.0								50
17	1986	10	22	04	20	44.9	38.42	116.14	3.0	25.5	0.22	0.58	0.25	11.0	3.00	0.16	127	78
18	1986	10	22	21	53	12.3	38.41	116.12	2.6	27.5	0.19	0.22	0.21	3.78	1.84	0.28	183	118
19	1986	10	25	13	43	40.2	38.38	116.12	2.0	12.0								45
20	1986	10	25	13	49	48.8	38.38	116.12	2.2	31.4	0.19	0.12	0.21	2.35	1.15	0.052	209	99
21	1986	10	27	11	44	12.0	38.45	116.12	2.6	23.5	0.32	0.12	0.36	3.00	0.28	0.022		71
22	1986	11	04	09	17	52.2	38.42	116.12	2.6	11.2								80
23	1986	11	04	09	21	07.4	38.42	116.13	2.1	13.0								84
24	1986	11	15	21	37	52.2	38.42	116.10	2.7	9.4								74
25	1986	11	15	22	16	23.0	38.40	116.12	3.2	12.0								
26	1986	11	19	22	23	36.5	38.42	116.13	2.6	11.2								102
27	1987	4	28	08	07	19.7	38.44	116.53	2.5	68.1								99
28	1987	6	02	14	26	45.0	38.47	116.04	4.5	11.2								72
29	1989	3	02	12	25	00.8	38.44	116.05	2.3	5.8								72

Note: M_L is the Chinese magnitude of body wave of local earthquakes.

program was initiated. Also, 4 months before the Xingtai M 6.0 earthquake on November 9, 1981, a production well spurted out a large volume of oil and water about 10^4 m^3 for 5 days. The borehole leak-outs in the oil field show that the direction of the maximum horizontal compressive stress is $60°N-80°E$, consistent with the P-axis direction in the North China plain. These observations indicate that the oil field is affected by the tectonic processes in the nearby Tangshan-Xingtai seismic zone in the North China plain. A few more sizeable faults pass through the oil field and strike NNE, consistent with the dominant direction of the seismic ruptures in North China. Furthermore, all the isoseismals in the oil field are elongated in the NNE or NE direction. The seismic intensity in the oil field has been progressively increasing from 1977 to 1987. In view of the above, we endeavored to understand the nature of the seismicity and rocks of the focal region.

Based on the reasons above, we have measured the focal parameters of those earthquakes and the medium characteristics. First applying the method of the least semi-period of P-wave's initial motions and the formulas given below. The focal parameters of 17 earthquakes include the dimensions of the fault surface, the seismic moments and the average interlocked distance. Next according to a single scattering mode, applying the relation of the peak amplitude of coda with time.

$$a = \frac{V_b}{\left(1 + \frac{\pi}{4}\frac{V_b}{C}\right)}(t_{2c}) \tag{1}$$

$$m_0 = \frac{u_{cm}}{(R_c^2)\frac{1}{2}\theta_c\frac{V_b}{a}} \cdot 4\pi\rho C^3 r \tag{2}$$

$$\Delta\sigma = \frac{7}{16}\cdot\frac{m_0}{a^3} \tag{3}$$

$$A\bar{u} = \frac{16}{7\pi}\cdot\frac{\Delta\sigma a}{u} \tag{4}$$

$$A(t) = \sqrt{8}M_0 I(f_p)1\frac{1}{4}-\frac{1}{2}\left|\frac{dt}{df_p}\right|^{-1/4}B(f_p)e^{-\pi f_p t/Q} \tag{5}$$

where,

a = fault size,
m_0 = seismic moment,
$\Delta\sigma$ = stress drop,
$\Delta\bar{u}$ = average dislocation,
A = coda amplitude,
Q = quality factor,
V_b = split velocity,
c = P velocity,

f = dominant frequency of coda,
I = amplification ratio of seismograph,
B = stimulation factor of coda.

We measured the maximum amplitude $(A_i(t))$ of each time window (t_i) and dominant frequency (f_i) from the place of maximum amplitude of S wave to the position at which the amplitude decreased to 2 mm with the time step of sampling 2 s. According to equations (2) and (3), the quality factor (Q) of 29 earthquakes was calculated. The stress drop varies from 0.2–3.0 bar with an average of 1.2 bar. The lower stress drop indicates that the average stress was not high and thus suggests that the earthquake occurred under low ambient stresses due to an increase in local fluid pressure. The Q value of coda varies from 45 to 118 and has an average value of 75. The low Q value is due to the fact that the focal region in Renqiu is very shallow, less than 5 km and reflects mainly the features of the upper crust, with the loose sediment of the Cenozoic Era and the fractured and porous rocks of the Paleozoic Era, where the coda is absorbed and scattered intensively. In other areas, the Q value of the upper crust is similar to the one in the Renqiu oil field. The low stress drop and the low quality factor reflect that the rocks of the upper crust in this oil field are brittle and of low strength. Hence, there cannot be a large accumulation of strain, implying that there is no possibility of the occurrence of large earthquakes in the Renqiu oil field due to water injection. However, small shocks will continue to occur, causing light damage. In view of this, people should not worry about induced earthquakes due to water injection.

The Unevenness of the Spatial Distribution of the Induced Earthquakes by Water Injection

The earthquakes induced by water injection are very unevenly distributed in the oil field. The northern part of the oil field (zone 1), characterized by the porosity of 5.95%, permeability of 0.04–0.59 darcy and the rate of seepage loss of 36.4%, exhibits high seismic activity, with the largest earthquake of magnitude 4.5. About 68% of the total induced earthquakes occurred in zone I, although it was the smallest area. The southern part of the oil field (zone III) with a porosity of 8.40–9.20%, permeability of 0.94–1.73% darcy and the rate of liquid loss of 85.7–88.2% is characterized by the low seismic activity, with the largest earthquake creating a magnitude of 2.5. Only 4% of the total earthquakes occurred in zone III. This demonstrates that frequency and magnitude of the induced earthquake are inversely related to the porosity and permeability of the reservoir rocks of the oil field (Fig. 6 and Table 2).

The results demonstrate that it is not easy to rupture the rocks with high porosity and therefore not easy to induce earthquakes, whereas in rocks with low

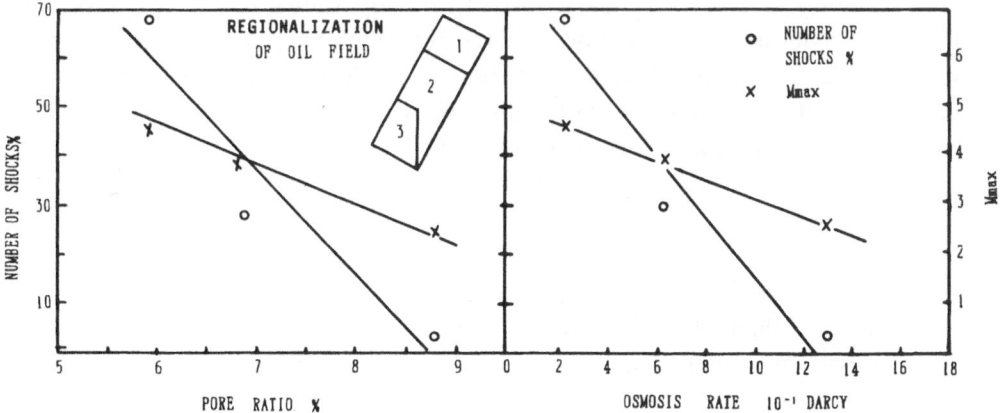

Figure 6
The relation between the earthquake activity and the porosity and permeability of the focal region rocks.

Table 2

Zone	Rock	Porosity %	Permeability darcy	Seepage loss of water %	Injection vol. km³/month	Seismic freq. %	M_{max} M_L
I	Limestone	5.95	0.04–0.59	36.4	<15	68	4.5
II	Dolomite	6.7–6.9	0.53–0.85	58.3–75.0	15–100	28	3.9
III	Dolomite	8.40–9.20	0.94–1.73	85.7–88.2	>100	4	2.5

porosity, high fluid pressure develops leading to high seismic activity. These characteristics will be valuable in the prediction of the frequency and scope of induced earthquakes due to water injection in an oil field.

Discussion and Conclusions

In the Renqiu oil field in the mainland of China, earthquakes induced by water injection are characterized by low stress drop, while the absorption and scattering of seismic waves in the rock medium is high. This implies that the earthquakes are caused by the brittle fracture of weaker rocks under low ambient stresses, due to a decrease in their strength resulting from the injection of water (high pore fluid pressures). These earthquakes are not large. However, their intensity in the epicentral area is very high reaching MM VI–VII. In only four of the 10 oil fields on the mainland of China, have earthquakes been induced by water injection. In the other six oil fields there exists a large range for aseismic slip which may destroy oil well bores resulting in severe economic loss. The study shows that earthquakes are induced by water injection into a deep oil reservoir with high fluid pressures and the pre-Paleozoic Era rocks of relatively higher strength. Contrastingly, only sliding takes place without earthquakes in the oil field, with shallow oil reservoir, low fluid pressures and the Cenozoic Era rocks of a ductile nature.

Acknowledgements

The authors thank Dr. H. K. Gupta and Dr. T. N. Gowd for their valuable suggestions in revising this paper.

REFERENCES

EVANS, D. M. (1967), *Man-made Earthquakes, A Progress Report*, Geotimes *12* (6), 19–20.
GUPTA, H. K., and RASTOGI, B. K., *Dams and Earthquakes* (Elsevier, Amsterdam 1976) 299 pp.
GENMO ZHAO, GUILING DIAO, and DEYUAN ZHANG (1981), *Seismicity in the Renqiu Oil Field and the Water Injection in Deep Well*, Researches on the Earthquakes in China, vol. 2.
GENMO ZHAO, and GANGSHEN YANG (1987), *Focal Stress Filled by Water Injection-induced Earthquakes and Characteristics of Coda Wave Attenuation*, Seismol. Geol. *12* (9), 301–310.
GUILING DIAO, DEYUAN ZHANG, and ZHAO GENMO (1982), *The Preliminary Researches on the Earthquakes Induced by Water Injection in Renqiu Oil Field*, Acta. Seismologica. Sinica in Northwest China *4* (3), 107–112.
HEALY, J. H., HAMILTON, R. M., and RAYLEIGH, C. B. (1970), *Earthquakes Induced by Fluid Injection and Explosion*, Tectonophys. *9*, 205–214.
NICHOLSON, C., ROELOFFS, E., and WESSON, R. L. (1988), *The Northwestern Ohio Earthquake of 31 January 1986: Was it Induced?*, BSSA *78* (1), 188–215.
RAYLEIGH, C. B., HEALY, J. H., and BREDEHOEFT, J. D. (1976), *An Experiment in Earthquake Control at Rangely, Colorado*, Science *191*, 1230–1237.
SIMPSON, D. W. (1986), *Triggered Earthquakes*, Ann. Rev. Earth. Planet. Sci. *14*, 21–42.

(Received May 13, 1994, revised/accepted February 2, 1995)

PAGEOPH, Vol. 145, No. 1 (1995)

0033–4553/95/010069–18$1.50 + 0.20/0

Earthquake Activity in the Aswan Region, Egypt

Mohamed Awad[1] and Megume Mizoue[2]

Abstract —The November 14, 1981 Aswan earthquake ($M_L = 5.7$), which was related to the impoundment of Lake Aswan, was followed by an extended sequence of earthquakes, and is investigated in this study. Earthquake data from June 1982 to late 1991, collected from the Aswan network, are classified into two sets on the basis of focal depth (i.e., shallow, or deeper than 10 km). It is determined that (a) shallow seismicity is characterized by swarm activity, whereas deep seismicity is characterized by a foreshock-main shock-aftershock sequence; (b) the b value is equal to 0.77 and 0.99 for the shallow and deep sequences, respectively; and (c) observations clearly indicate that the temporal variations of shallow seismic activity were associated with a high rate of water-level fluctuation in Lake Aswan; a correlation with the deeper earthquake sequence, however, is not evident. These features, as well as the tomographic characteristics of the Aswan region (Awad and Mizoue, this issue), imply that the Aswan seismic activity must be regarded as consisting of two distinct earthquake groups.

We also relocated the largest 500 earthquakes to determine their seismotectonic characteristics. The results reveal that the epicenters are well distributed along four fault segments, which constitute a conjugate pattern in the region. Moreover, fault-plane solutions are determined for several earthquakes selected from each segment, which, along with the 14 November 1981 main shock, demonstrate a prominent E-W compressional stress.

Key words: Lake Aswan, tectonic setting, seismicity.

Introduction

The Aswan seismic region is located 70 km southwest of the Aswan High Dam, beneath a western embayment of Lake Aswan (Fig. 1). The Aswan High Dam was constructed in early 1964 across the Nile River in the Aswan prefecture of Egypt, and is a unique structure among irrigation and electric power projects in the world. As a consequence of the construction of the Aswan High Dam, the world's second largest reservoir has developed in southern Egypt. The reservoir level rose progressively during the annual irrigation cycles until a maximum level of 177.48 m was achieved in November 1978. The ground water extension of the reservoir in the Nubian sandstone formation was modeled by Evans *et al.* (1991). Simpson *et al.*

---

[1] National Research Institute of Astronomy and Geophysics, Helwan, Cairo, Egypt.
[2] Earthquake Research Institute, Tokyo University, 1-1 Yayoi 1-Chome, Bunkyo-ku, Tokyo 113, Japan.

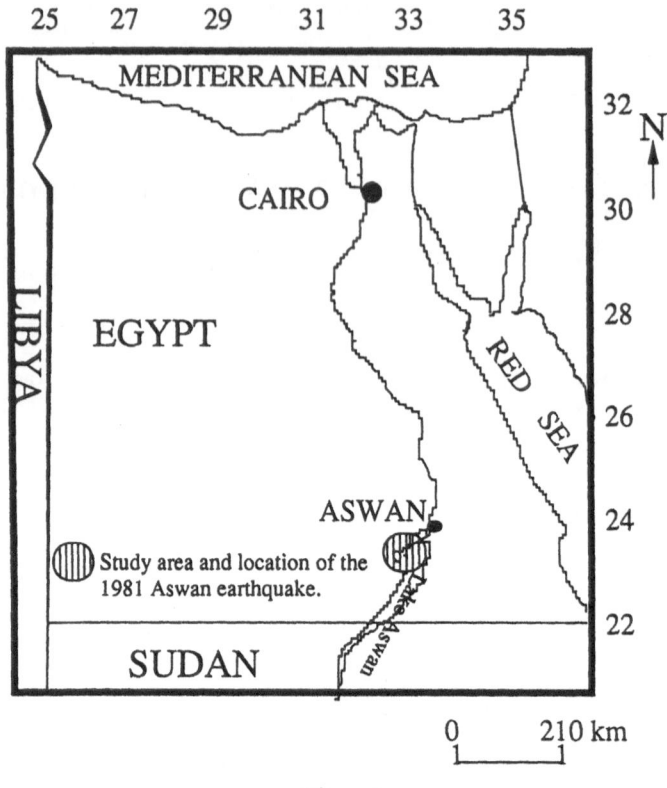

Figure 1
Study area.

(1992) investigated the variations in the lateral extent of the reservoir and the seismicity-water level relationship.

The tectonic setting in the Aswan region is dominated by regional uplift and the presence of two fault systems trending in the E-W and N-S directions (Fig. 2) (ISSAWI, 1971, 1982, 1987). The Kalabsha fault is well-defined in aerial photographs and is characterized by enchelon folds. The fault consists of several segments with a total length of 300 km in the E-W direction. It is located on the western side of Lake Aswan beneath the Kalabsha embayment (Fig. 2). Geologic evidence indicates that the fault is associated with right-lateral movement (KEBEASY et al., 1987; VYSKOCIL et al., 1987, 1989). Approximately 12 km north of the Kalabsha fault is the Seiyal fault, which is also characterized by an E-W trend. The Kalabsha and Seiyal faults constitute a graben structure occupied by the Kalabsha embayment (Fig. 2) (ISSAWI, 1982). The N-S fault trend consists of several fault segments (Fig. 2).

Prior to the November 14, 1981 earthquake, which occurred along the Kalabsha fault (KEBEASY et al., 1982), information on seismicity in the Aswan region is limited due to the lack of local seismograph stations at that time. A few small

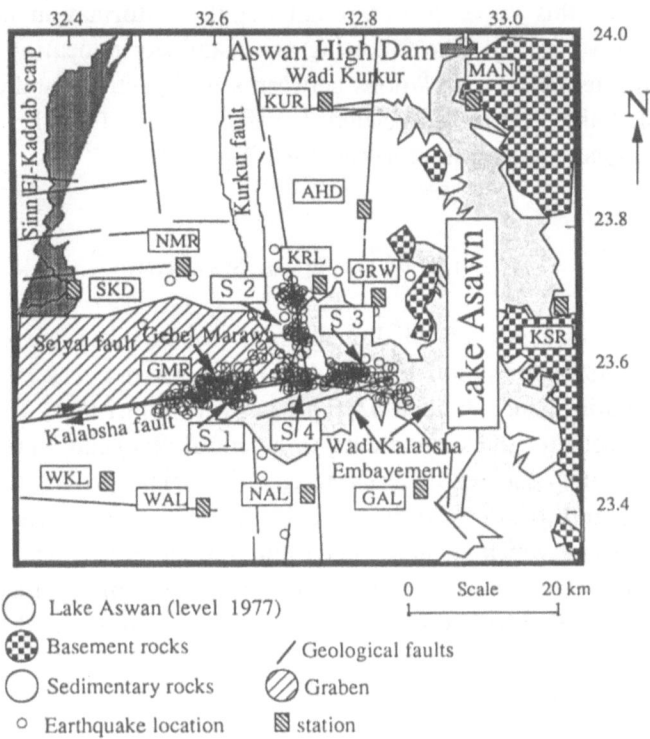

Figure 2
Location map of the Aswan network stations, tectonic setting and epicentral distribution of the relocated earthquakes. S1 to S4 are fault segments described in the focal mechanism solutions.

shocks ($M < 3$) were detected in the southernmost part of Lake Aswan in the town of Abu Simple in January 1981 during a reconnaissance survey to measure the back-ground noise level (GIBOWICZ et al., 1982). Seismograms of a short-period seismic station located at Abu Simple indicate that few microearthquakes can be located within the Kalabsha region of the Aswan seismic area (KEBEASY et al., 1982). The November 1981 earthquake was preceded by three foreshocks; consisting of two events on November 9 ($M_L = 3.6$ and 4.2) and one event on November 11 ($M_L = 4.5$), and followed by an extended sequence of microearthquakes (KEBEASY et al., 1982). The focal depth of the main shock was determined to be 20 km. On January 2, 1982, two successive earthquakes occurred at shallow depth (10 km).

The crustal structure in this region was estimated using an explosion experiment in 1986 (KEBEASY et al., 1992), and the three-dimensional P-wave velocity structure was recently inverted from the local earthquake travel time data (AWAD and MIZOUE, this issue). Analysis of the crustal deformation reveals substantial subsidence in the southern part of the Aswan seismic region (VYSKOCIL et al., 1987, 1989).

The purpose of this paper is to present detailed information concerning the nature of Aswan seismicity. The following are examined: relocation of the largest 500 earthquakes, temporal correlations between seismic activity and changes in the water level of Lake Aswan, associated b value and earthquake patterns, and composite fault-plane solutions of selected earthquakes.

Analysis of the Aswan Seismicity

An extended sequence of small-magnitude earthquakes ($M < 5$) was well recorded in the Aswan seismic region with a radio-link network that has been in continuous operation since June 1982. This network consists of 13 field stations (Fig. 2) which telemeter the seismic data to a central recording unit in the Aswan Seismological Center, located in Aswan city. Seismic signals are recorded on magnetic tapes, and drum recorders are available for only five stations. The preliminary analysis, which includes determination of earthquake locations and magnitudes, is performed by the staff of the Aswan Seismological Center which is associated with the National Research Institute of Astronomy and Geophysics at Helwan in Egypt.

Earthquake Relocation

We relocate the largest 500 earthquakes of Aswan seismic activity using two inversion programs; one was developed by HIRATA and MATSU'RA (1987) and the second is the three-dimensional Thurber's Program (THURBER, 1983). The seismograms were obtained from an analog recorder which is available in the Aswan Seismological Center. The P- and S-arrival time data were measured to an accuracy of 0.02 and 0.1 s, respectively. Station corrections are determined for all observation sites from the P and S residuals of an initial hypocentral determination by dividing the sum of the residuals by the number of events per station. The assumed model of the local crustal structure is that determined in the three-dimensional P-wave velocity study (AWAD and MIZOUE, this issue) and is shown in Figure 3, whereas values of contour lines are the P-wave velocity (km/s) along the seismically active zone of Aswan crust. Hypocentral parameters of the relocated earthquakes, which are characterized by less than 1.5 and 2.0 km error in the epicentral coordinates and the focal depth, respectively, are well distributed along four geological faults in the study region (Fig. 2). Three fault segments (S1, S4 and S3 in Fig. 2), collectively referred to as the Kalabsha fault, extend in the E-W direction, whereas the fourth (S2) trends N-S and coincides with the southern portion of the Kurkur fault. The N-S and E-W cross sections are utilized to illustrate the depth distribution of the hypocenters of the Aswan seismic activity including the relocated earthquakes (Fig. 3). This figure demonstrates the presence of two distinct seismic zones, which are

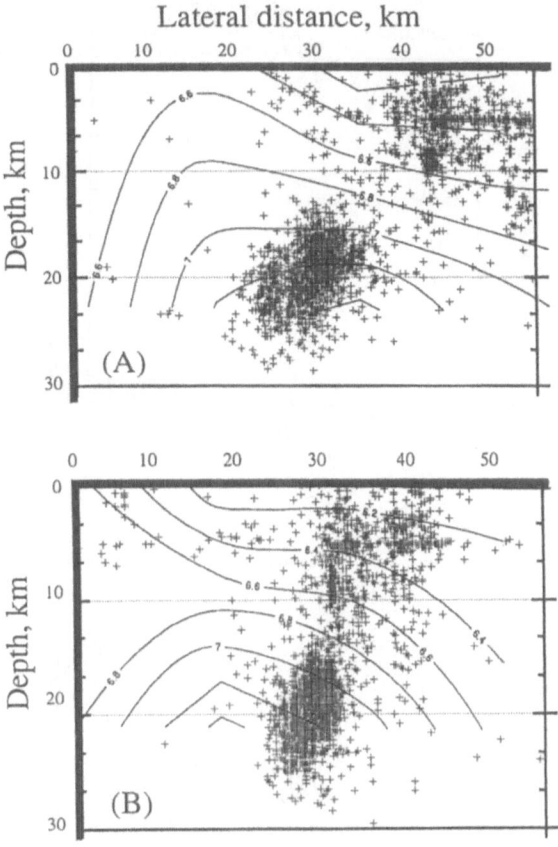

Figure 3

Depth distribution of Aswan seismicity including the relocated earthquakes. Values of the contour lines are the *P*-wave velocity (km/s). (A) N-S cross section. (B) E-W cross section.

located within regions of low- and high-velocity anomalies in the shallow and deep parts of the Aswan crust, respectively. The deeper zone is an oblique segment dipping in the southwest direction. The shallow earthquakes are generally scattered (Fig. 3).

Seismicity and Water-level Variations in the Aswan Lake

Previous studies of the temporal variations of Aswan seismicity and water-level variations in Lake Aswan were performed by KEBEASY *et al.* (1982, 1987) and SIMPSON *et al.* (1992). The summary of seismicity and water-level correlations may be found in SIMPSON *et al.* (1992), which addresses the time period from 1964 to 1989 (Fig. 4). In their correlation, the seismic activity in the Aswan region is considered to represent one uniform set of earthquakes. Our results indicate,

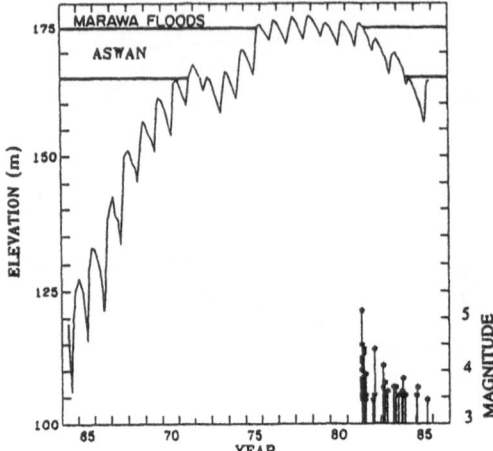

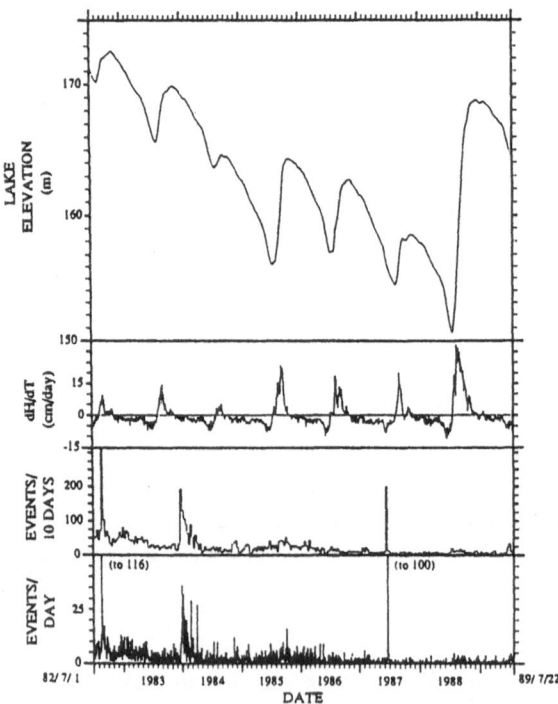

Figure 4
Lake Aswan water-level variations and Aswan seismic activity, after SIMPSON *et al.* (1992).

however, that this activity occurred within two distinct depth intervals, one from 4 to 8 km and another from 14 to 26 km. In this study, the shallow and deep earthquake sequences are independently correlated with water-level variations. In Figure 5 we plot the daily number of earthquakes in each group with respect to the water-level changes in the lake, as well as the daily average of water-level fluctuations. The number of earthquakes during the activities denoted I and II in Figure 5 can be seen in Figure 7. Note that since June 1982, the Aswan seismicity is limited to a magnitude level of less than 4.9.

Water-level measurements in the lake are determined four times per day by the High Dam Authority in Aswan. We utilize the daily average of these measurements to determine the correlation with the temporal seismicity variations. The rate of the water-level change (daily fluctuation) is defined by the difference in water height between two successive days. The Aswan region is characterized by an arid climite, and the water level in the lake is dominated by a combination of seasonal changes and irrigation requirements.

Deep Earthquake Sequence

The largest earthquake to have occurred since the early aftershock sequence in November-December 1981, was a magnitude 4.9 event on 20 August 1982. This event, and the associated sequence, occurred in the deep zone beneath Gebel Marawa (Fig. 3), in the same region in which the November 14, 1981 main shock was located (ISC Bulletin). The associated activity decreased by late February 1983 (Fig. 5). From November 22, 1982 to the end of February 1983 (Fig. 5) there was a small burst of seismic activity characterized by low-magnitude events, of which the largest occurred on February 24, 1983. The seasonal minimum water level occurred on 11 August 1982, 9 days before the August 20 earthquake. The annual peak in water level occurred on November 13, 9 days before the slight increase in deep seismicity began and 103 days before the occurrence of the February 24, 1983 earthquake. During this period, daily average rates of water-level variations are not correlated with the deep earthquake sequence, but they represent the correlation with the shallow earthquake group as is described below. It should be noted that the water-level in Lake Aswan has been affected by the recent drought in Africa. This is apparent in an annual decrease that persisted until the end of 1988, when a sharp increase occurred (Fig. 4). The annual change in water level in 1984 was considerably less than in previous years. Deep earthquakes have been rare since February 1983.

Shallow Earthquake Sequence

Among the shallow seismic activity, that which occurred during December 1983 to March 1984 exhibits an interesting correlation with the daily rate of water-level change in Lake Aswan (Fig. 5). During this period, the region experienced a burst of shallow seismic activity beginning at the end of 1983. This activity was concentrated at shallow depth (2 to 7 km) beneath the area covered by water in the main course of the Lake and Kalabsha embayment (northeast of Gebel Marawa), and is characterized by small-magnitude earthquakes ($M < 3.3$). The water level of the lake fluctuated considerably due to irrigation requirements during the period from December 20, 1983 to the end of March 1984, and reached a "peak-to-peak" amplitude of 14 cm between January 1 and 2, 1984. The seasonal maximum was attained on December 10, 1983. Records of the daily rate of the water-level change and seismic activity during this time period reveal a clear correlation (Fig. 5).

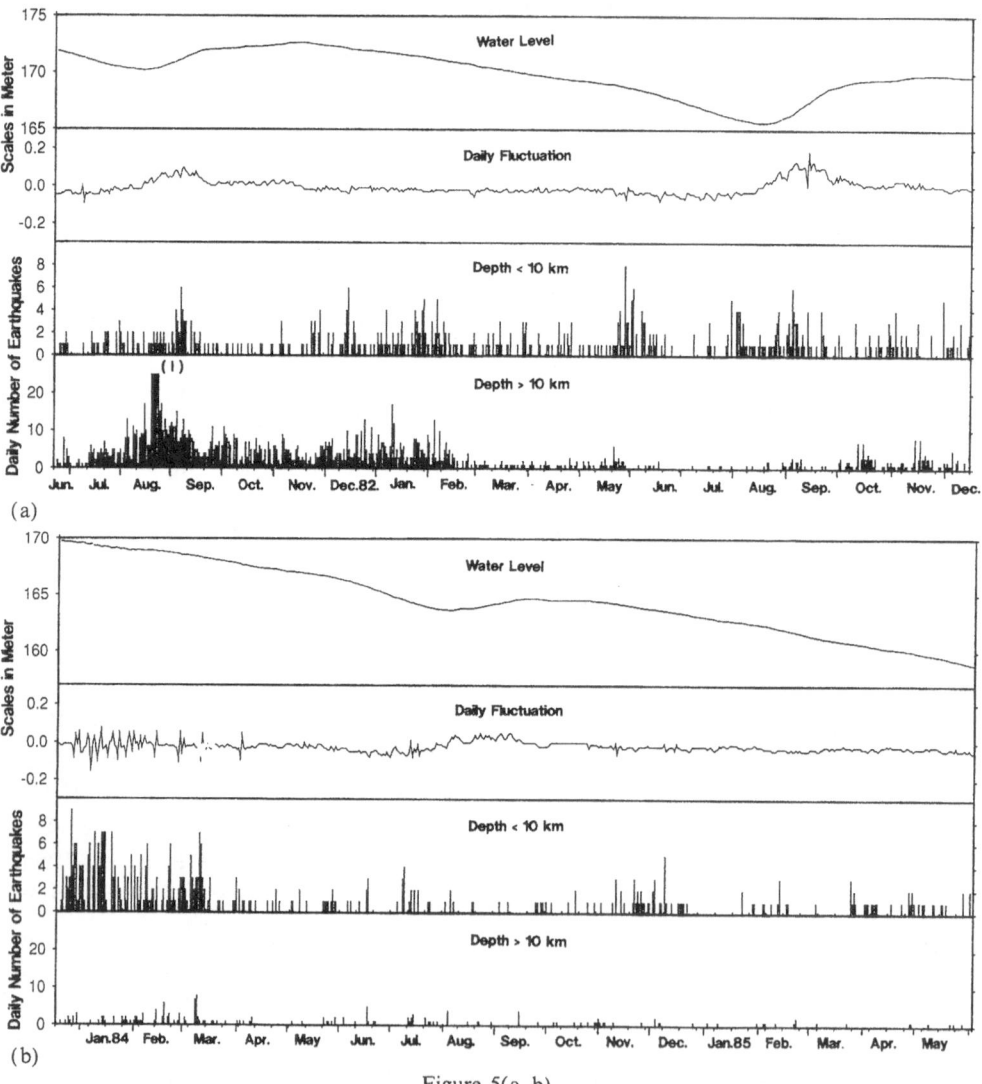

Figure 5(a, b)

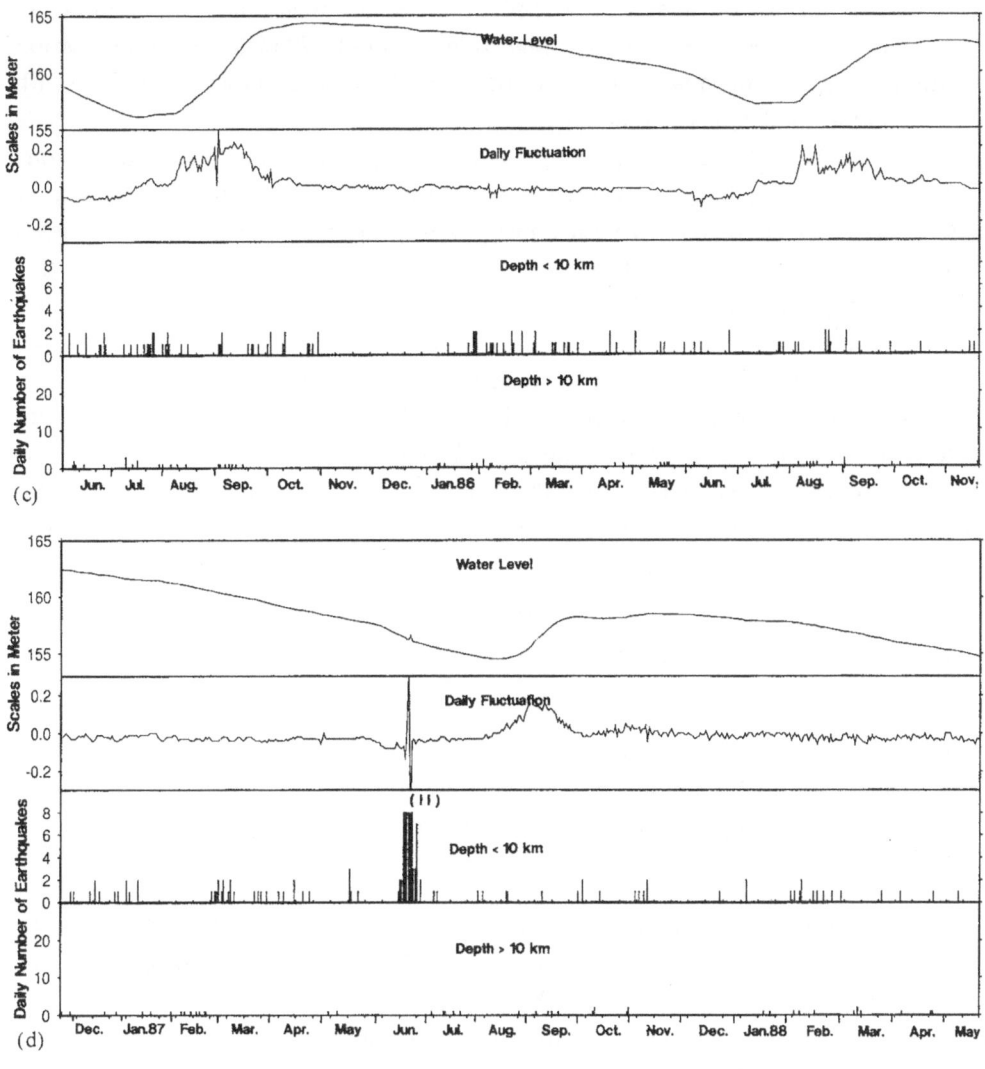

Figure 5(c, d)
Figure 5
Earthquake time distributions in the Aswan region and water-level variations in Lake Aswan in the period from June 20, 1982 to the end of May 1988.

From April 1984 to June 14, 1987, although the shallow events are small in number and of low magnitude, one can find an earthquake at any time when the fluctuation in the water level was large. This observation can be noted throughout the histogram (Fig. 5). The maximum "peak-to-peak" amplitude of fluctuation during this period was 26 cm between October 12 and 13, 1985, and was not associated with an increase in seismic activity, whereas only five earthquakes had occurred. This may be because the water level was influenced by the African drought, where there is an enormous difference between the maximum seasonal water levels during the first and second time periods mentioned above (Fig. 5). Note that when the water level was high (e.g., around 170 m in the earlier period of the histogram), the small amplitudes of the daily fluctuation are associated with a remarkable shallow seismic activity.

On June 15, 1987, a short, abrupt, intense swarm began in the shallow zone between 4 and 8 km depth. The earthquake frequency was very high during June 17 and 19 and was characterized by a typical swarm pattern, which will be discussed below. The most interesting feature of this swarm activity is that it followed a sudden change in water level, which had occurred between June 15 and 16 and between June 18 and 19. The amplitude of the daily fluctuations "peak-to-peak" records a 21 cm and 1 m change between the above-mentioned days, respectively. This observation and the correlations observed during the histogram period indicate that there must be a relationship between the shallow seismic activity and the daily fluctuation of water level in Lake Aswan. It should be mentioned also, that there has been no dramatic increase in the shallow seismicity associating some recorded changes in the daily water level.

b Value and Depth Distribution

We determined the frequency-magnitude relationship (*b* value) in GUTENBERG and RICHTER (1956) for both the shallow and deep seismic activity, which occurred during the period of June 20, 1982 to the end of 1990. In this study, earthquake magnitudes are determined using the time-duration empirical formula

$$M = -0.87 + 2 \log T + 0.0035D$$

where *M* is the magnitude, *T* is the *F* − *P* time duration in seconds and *D* is the epicentral distance in kilometers (LEE *et al.*, 1972). Using a simple program of the least square method, it is obtained that the *b* value equals 0.77 and 0.99 for the shallow and deep seismic activity within the above-mentioned period, respectively.

The number of earthquakes with respect to the depth (in increments of 5 km) is shown in Figure 6, which illustrates the concentrated zones in the Aswan seismic region. The selection of 5 km for each class is in accordance with the average error of the focal depth estimations of the bulletins used. This depth distribution is shown in five histograms, the upper most is drawn on the Egyptian Geological Survey data, whereas the remaining histograms are based on the Aswan network data. The hypocentral parameters of both data sets were determined using the Hypoinverse program (KLEIN, 1978) (see BOULOUS *et al.*, 1987 and the ASWAN SEISMOLOGICAL CENTER Bulletin). These histograms indicate that the seismic activity was concentrated in two depth intervals, from 5 to 10 km and from 15 to 25 km, and that the activity in this time period was initially concentrated in the deeper zone and migrated upward to shallower depths during the later stages.

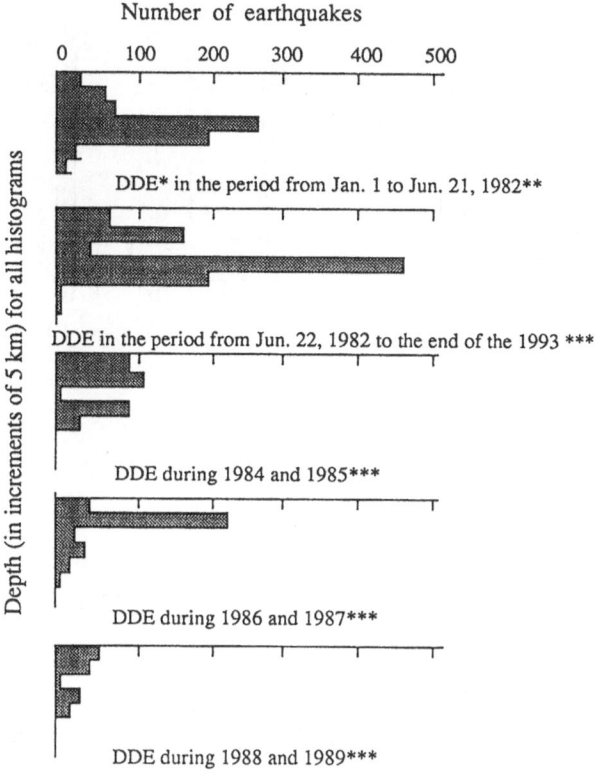

Figure 6

Histograms of earthquake depth-distributions in the Aswan region. *DDE means depth distribution of earthquakes, ** data of the Egyptian Geological Survey, *** data of the Aswan seismic network.

Earthquake Patterns

Throughout the extended sequence of Aswan seismic activity there are only two well-recognized peaks in activity. The first occurred in August 1982 within the deeper seismic zone (i.e., at depths of 18 to 26 km). The main shock ($M = 4.9$), took place on August 20, and was followed by an exponential decay of aftershocks (Fig. 7). On August 20, 113 earthquakes, with magnitudes less than 3.0, occurred. The main shock of this sequence was approximately two units larger than the most powerful aftershock, and was preceded by few foreshocks (Fig. 7).

The second burst occurred at shallow depth (i.e., from 4 to 9 km) in June 1987 and was associated with a sudden change in the water level in Lake Aswan. The rate of earthquake occurrence was high on the days of June 17 and 19 (Fig. 7). During this burst, six events, of magnitude between 3.0 and 3.4, occurred. The time-magnitude distribution of earthquakes related to this burst exhibits a typical swarm sequence (MOGI, 1967).

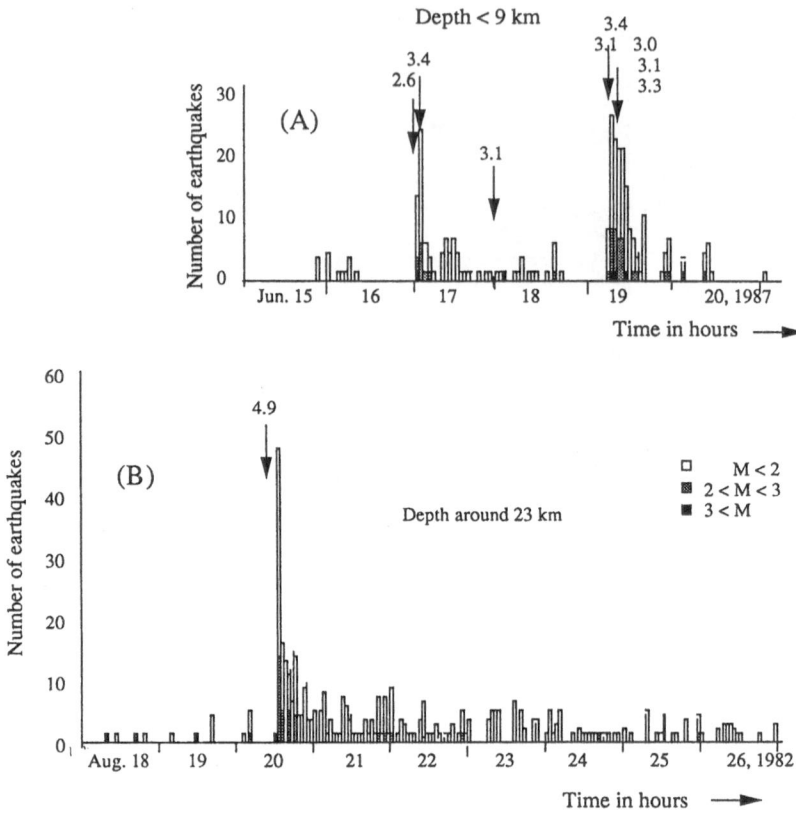

Figure 7
Time-magnitude distribution of two peaks in the earthquake activity in the Aswan region. (A) The June 1987 swarm. (B) The August 1982 activity.

Focal Mechanism Analysis

Composite fault-plane solutions are determined for 97 earthquakes selected from the relocated data using P-wave first motions. These earthquakes are of small magnitude ($M < 4.5$) and distributed along four faults denoted as S1 to S4 in Fig. 2. All first-motion readings were obtained from thermal-paper seismograms recorded by the local Aswan telemetered network (epicentral distance < 60 km). Takeoff angles were computed using the velocity model in AWAD and MIZOUE (this issue) and the earthquake locations. Since the quality of most of the locations is excellent and the velocity model is well-known, the computed takeoff angles are reliable. First-motion plots for all 97 earthquakes are obtained utilizing a computer program written by MAKI from the Japan Meteorological Agency. All projections were performed on a lower-hemisphere net of an equal area projection (Schmidt net). The composite fault-plane solutions for the four segments are determined by

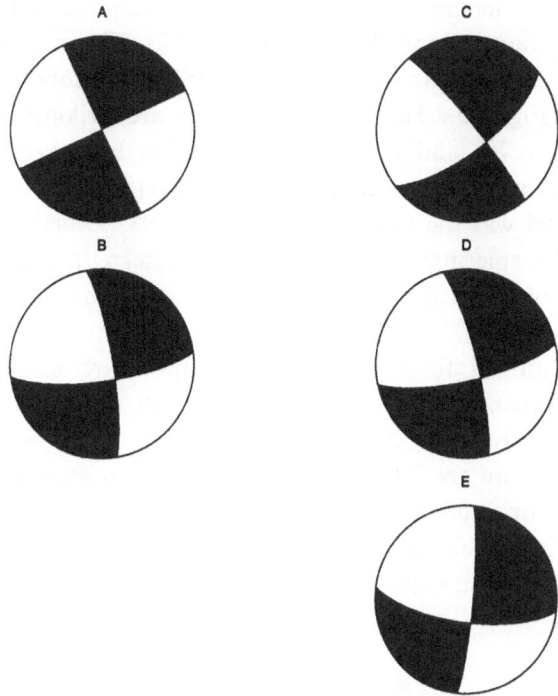

Figure 8
Focal mechanism analyses for the Aswan seismic activity. A solution for the November 14, 1981 main shock; see also KEBEASY *et al.* (1987). B, C, D and E solutions for the fault segment S1, S2, S3 and S4 of Fig. 2, respectively. Projection on a lower hemisphere.

a visual fit, where orthogonal planes were added to separate the first-motion data into quadrants of compression and dilatation. The resulting composite fault-plane solution diagram for each segment is shown in Figures 8B,C,D,E. The solution shown in Figure 8A is for November 14, 1981, based upon the teleseismic data and it agrees well with that obtained by KEBEASY *et al.* (1982, 1987).

Fault segment S1 is located beneath the Gebel Marawa along the section of the Kalabsha fault involved in the November 14, 1981 main shock, and is characterized by strike-slip motion (Fig. 8A). This segment contains earthquakes from the August 1981 burst. The composite mechanism solution (Fig. 8B) of this segment represents the normal-fault component; the two nodal planes are striking 80° and 170° from the north with dips of 70° and 80°, respectively. The nodal plane striking 80° is favored because this strike is nearly parallel to the Kalabsha fault.

Fault segment S2 is located on the southern portion of Kurkur fault, which is about 25 km NE of Gebel Marawa (Fig. 2). Earthquake epicenters of this segment are situated in a nearly N-S direction. The solution presented in Figure 8C indicates a strike-slip fault. The two nodal planes strike 50° and 140° from the north with dips 65°and 83°, respectively. As the Kurkur fault trends N-S, the fault plane is assigned to the nodal plane of strike 120° from the north.

Fault segment S3 includes the eastern part of the Kalabsha fault, which is disrupted by other faults along this portion (Fig. 2). The earthquakes trend in a nearly ENE-WSW direction and were characterized by shallow foci. The solution is a strike-slip fault (Fig. 8D). The two nodal planes are striking 75° and 165° from the north with dips 70° and 80°, respectively. The ENE-WSW nodal plane is favored as the fault plane (in relation to the fault setting).

Fault segment S4 was the location of the June 1987 swarm, which occurred at a shallow depth. The epicenters are located at the western tip of the E-segment (Fig. 2). The two nodal planes strike at 5° and 95° from the north, with dips of 83° and 65°, respectively (Fig. 8E).

The above-mentioned focal mechanism solutions are nearly similar to one another, the nodal plane that is assigned as the fault is determined based on the local geologic structure (Fig. 2), and the distribution of epicenters along each segment. Although there are a few inconsistent events in each diagram, these may be attributed to minor trends within each source region. An approximately NE-SW compressional axis can be inferred from these solutions, as well as the focal-mechanism solution of the main shock (Fig. 8A).

Results and Discussion

We propose that the Aswan seismic activity should be regarded as two earthquake groups rather than one uniform set. The Aswan region is characterized by a heterogeneous crust with the presence of low- and high-velocity zones in the shallow and deep portions of the crust, respectively. Most of the shallow earthquakes are located within the low-velocity zone, and the deeper events coincide with the high-velocity anomaly. The two classes of events are therefore presumably associated with different environments, which constitute an important reason for the classification of Aswan seismicity into shallow and deep earthquake groups. We present different analyses for the Aswan earthquake activity during the period of June 1982 to the end of 1991. These analyses demonstrate several other characteristic features, which also support the classification proposed in this paper. One such feature is the difference in the earthquake patterns and the constant b value. The shallow earthquake group is characterized by a small b value (0.77) and a swarm pattern (Fig. 7). Whereas the deeper seismic zone is characterized by a foreshock-main shock-aftershock pattern and a normal b value of 0.99. These two observations were used to classify the earthquake activity into different types by Mogi (1962), who pointed out that they are related to the medium characteristics and/or the applied stresses. Mogi (1967) discussed the relationship between these two aspects and the different seismic active regions in and around Japan, and has shown a good correlation. In this context, application to the Aswan seismicity illustrates that the shallow and deep earthquake groups are associated with a different mode of strain release, which is related to the earthquake pattern and the constant b value. In other words, each earthquake group occurred under different physical conditions and/or mode of applied stress.

The second characteristic feature is the correlation between shallow seismicity and water-level variations in the reservoir, which is not recognized for the deeper earthquake sequence. In fact, the reservoir-induced seismicity is often characterized by shallow focus events, which occur in a short-term period after water-level changes in the reservoir (SIMPSON and NEGMATULLAEV, 1981; TAPPOZADA and MORRISON, 1982; GUPTA 1983). Most of the shallow earthquakes in the Aswan region exhibit a relationship to water-level fluctuations (Fig. 5). The question that arises in this context is whether this state of coexistence between the shallow earthquakes and the water-level fluctuation has developed since the filling began in 1964 or just after the occurrence of the deeper earthquake sequence. The answer is quite complicated due to the lack of local stations within the Aswan seismic region prior to January 1982. The nearest station was irregularly operated in the town of Abu Simple, which is located at a distance of about 280 km from the seismically-active region. Several small-magnitude ($M_L < 4.0$) earthquakes were recorded by this station and may be located within the Aswan seismic region (KEBEASY et al., 1982), however, their hypocentral parameters cannot be reliably established. On the other hand, the earthquake frequency-depth distribution shown in the histograms of Figure 6 indicates that the seismic activity has migrated up to the shallower zone in the later stages from the initially deeper seismic zone in the Aswan crust. One possible interpretation is that the shallow zone has been influenced by the occurrence of the deeper earthquake sequence and became weaker and therefore more easily influenced by the reservoir controlling factors of earthquake generation. In this scenario the November 14, 1981 event (20 km depth) bears no relation to the existence of the reservoir. Of course, there might be a relationship between the reservoir effect and the deeper events, but a plausible mechanism has not yet been demonstrated. The interesting point in this correlation is that the classification of earthquakes can be modeled under the assumption of events which are related to the reservoir fluctuation and those which are not, because the correlation does not characterize the deeper seismic zone in the study region. In fact, the Aswan seismic activity deviates from the classical concept of reservoir-induced seismicity because the presence of the deeper seismic zone, as well as the long time delay between the start of filling and the occurrence of the November 14, 1981 main shock, are anomalous (SIMPSON et al., 1992; KNOLL, 1992).

Epicenters of the relocated earthquakes (Fig. 2) are well distributed along only four segments of the faults appearing in the geological map of this region (Fig. 2). Three segments are characterized by a shallow seismicity and extend beneath the Kalabsha embayment of the reservoir. These segments are within the region in which a decrease in the seismic velocity occurs (AWAD and MIZOUE, this issue). Such a well-defined correlation in the space domain, distribution of epicenters, tectonic setting and the presence of a low-velocity zone, provides information about the nature of the seismic-velocity reduction in this zone. The faults represent weak zones where fluids may presumably penetrate and produce a pressure, which may cause a reduction in the seismic velocity (RALEIGH and EVERNDEN, 1981). Presence of the microcracks in the Aswan region is demonstrated by the occurrence of many, very small-magnitude shocks ($M < 1.5$). Some of these cracks may trend in a different direction with respect to the major segment trend. This may also explain our observation of the few events that are inconsistent with the general solution in the composite fault-plane analysis (Fig. 8), which can be seen in Fig. 19, page 48 in the Ph.D. thesis of AWAD (1994).

The Aswan region is dominated by a succession of tectonic, erosional and igneous events which produced the present complex of Aswan granite and other rock types exposed at the surface within the Nubian sandstone plain (ISSAWI, 1971, 1978, 1982, 1987 and EL SHAZLY, 1977). In the deeper seismic zone, the relocated events are distributed along the western segment of the Kalabsha fault, which extends to the west from a location that is dissected by the E-W and N-S fault trends and is situated just off the end of the Kalabsha embayment. In this joint region the S4 segment appears to be an example of a minor triple junction between the epicenters of the other three segments S1, S2 and S3 (Fig. 2). Hypocenters of the deeper segment are distributed in a concentrated oblique zone, which extends almost along the total thickness of the lower-crustal layer (Fig. 3) and dips towards the southwest. The zone of earthquakes in this segment coincides with the deeper, high seismic-velocity anomaly. The high-velocity anomaly may be related to the presence of an intrusive body in the lower crustal layer (AWAD and MIZOUE, this issue). If this is the case, the intrusive activity may produce a vertical pressure, which would explain the normal-faulting component obtained in the composite fault-plane solutions (Fig. 8).

The composite fault-plane analyses illustrate that strike-slip faulting is the predominant mechanism along the four active segments in the Aswan region. Figure 8 illustrates that the hanging walls represent the southern blocks for the three segments of the Kalabsha fault. For the N-S segment, the eastern block constitutes the hanging wall. These results are consistent with crustal deformation measurements in the Aswan region, where subsidence is observed in the southern part of the Kalabsha fault and along the eastern side of the Kurkur fault (VYSKOCIL et al., 1989). The focal mechanism analyses also indicate that the Aswan region is subjected to a nearly NE-SW compressional stress, which is correlated with the regional tectonics in southern Egypt. The southern part is regionally affected by a compressional stress field related to the spreading center of the Red Sea (MCKENZIE et al., 1970; BEN MENAHEM et al., 1976). A detailed investigation of the regional tectonics is warranted due to the lack of correlation of the reservoir and dynamics and the deeper seismic zone.

Conclusion

The extended sequence of earthquakes in the Aswan region occurred along four fault segments, three of which extend in the E-W direction and are associated with the Kalabsha fault zone. The fourth segment exhibits a N-S trend and coincides with the southern part of the Kurkur fault. The extensions of these active segments constitute a net of conjugate faults. The results of the focal-mechanism solutions are well correlated with the geologic interpretation of the local tectonics and geodetic measurements. There are several seismicity features that dictate that the earthquake sequence be classified into shallow and deep groups: (1) The shallow earthquakes are well correlated with the daily average rate of water-level variations in the lake, whereas this is not true for the deeper seismic activity; (2) although most of the shallow earthquakes coincide with a low seismic-velocity zone, the deeper events concide with a high-velocity zone; and (3) the shallow earthquakes are characterized by a small b value (0.77) and a swarm pattern, however the deeper earthquakes are characterized by a normal b value (0.99) and a foreshock-main shock-aftershock pattern.

Acknowledgments

We acknowledge the efforts of Profs. R. M. Kebeasy and D. Simpson in establishing the Aswan network. We are grateful to Prof. Dr. H. K. Gupta for his kind response to examine this paper. We thank also Drs. S. Mueller and K. Tsukada for their kindly help during the manuscript preparation.

REFERENCES

AWAD, E. M. (1994), *Investigation of the Tectonic Setting, Seismic Activity and Crustal Deformation in Aswan Seismic Region, Egypt*, Ph.D. Thesis, Tokyo University.

AWAD, M., and MIZOUE, M. (1995), *Tomographic Inversion for Three-dimensional P-wave Velocity Structure of the Aswan Region, Egypt* (this issue).

BEN MENAHEM, A., NUR, A., and VERED, M. (1976), *Tectonic, Seismicity and Structure of Afro-Eurasian Junction — the Breaking of an Incoherent Plate*, Phys. Earth. Planet. Int. *12*, 1–50.

BOULOS, F. K., MORGAN, P., and TOPPOZADA, T. (1987), *Microearthquake Studies in Egypt Carried out by the Geological Survey of Egypt*, J. Geodyn. *7*, 227–249.

EL SHAZLY, E. M. *The Geology of the Egyptian Region. The Ocean Basin and Margins*, vol. 4A (Plenum 1977) 379–444.

EVANS, K., BEAVAN, J., SIMPSON, D., and MOUSA, S. (1991), *Estimating Aquifer Parameters from Analysis of Forced Fluction in Well Level: An Example from the Nubian Formation Near Aswan, Egypt*, Successive three papers, J. Geophys. Res. *96* (B7), (12), 127, 189.

GIBOWICZ, S. J., DROSTE, Z., KEBEASY, R. M., IBRAHIM, I. M., and ALBERT, R. N. H. (1982), *A Microearthquake Survey in Abu-Simple Area in Egypt*, Engin. Geol. *19*, 95–109.

GUPTA, H. K. (1983), *Induced Seismicity Hazard Mitigation Through Water Level Manipulation at Koyna, India. A Suggestion*, Bull. Seismol. Soc. Am. *73*, 679–682.

GUTTENBERG, B., and RICHTER, C. F. (1956), *Magnitude and Energy of Earthquakes*, Ann. Geofis. *9*, 1–15.

HIRATA, N., and MATSU'RA, M. (1987), *Maximum-likelihood Estimation of Hypocenter with Origin Time Estimated Using Nonlinear Inversion Technique*, Phys. Earth. Plant. Int. *47*, 50–61.

ISSAWI, B. (1971), *Geology of the Darb El-Arbain, Western Desert*. Annals Geolog. Survey of Egypt.

ISSAWI, B. (1978), *Geology of Nubia West Area. Western Desert*. Annals Geolog. Survey of Egypt.

ISSAWI, B. (1982), *Geology of the Southwestern Desert of Egypt*. Annals Geolog. Survey of Egypt.

ISSAWI, B. (1987), *Geology of the Aswan Desert*. Annals Geolog. Survey of Egypt.

KEBEASY, R. M., MAAMOUN, M., and IBRAHIM, E. M. (1982), *Aswan Lake Induced Earthquakes*, Bull. Inter. Inst. Seismol. and Earthq. Engin. Tokyo *19*.

KEBEASY, R. M., MAAMOUN, M., IBRAHIM, E. M., MEGAHED, A., SIMPSON, D. W., and LEITH, W. S. (1987), *Earthquake Studies at Aswan Resevoir*, J. Geodyn. *7*, 173–193.

KEBEASY, R. M., BAYOUMI, A. I., and GHARIB, A. A. (1992), *Crustal Structure Modelling for the Northern Part of the Aswan Lake Area Using Seismic Waves Generated by Explosions and Local Earthquakes*, J. Geodyn. *14*, 1–24.

KLEIN (1978).

KNOLL, P., *The dynamic excess pore pressure concept — A new possible fracture mechanism for fluid-induced seismic events*. In *Induced Seismicity* (Peter Knoll, ed.) (Central Institute for Physics of the Earth, Potsdam 1992) pp. 275–286.

LEE, W., and LAHR, J. C. (1972), *A Computer Program for Determining the Hypoceuter, Magnitude and First Motion Pattern of Local Earthquake. Revision of HYPOTI*, USAS, open file report., 100 pp.

LEE, W., *Microearthquake Networks* (Academic Press, New York 1982).

MCKENZIE, D. P., DAVIES, D., and MOLNAR, P. (1970), *The Plate Tectonics of the Red Sea and East Africa*, Nature *226*, 243–248.

MOGI, K. (1962), *On the Time Distribution of Aftershocks Accompanying the Recent Major Earthquakes in and near Japan*, Bull. Earthq. Res. Inst., Tokyo Univ. *40*, 107–124.

MOGI, K. (1967), *Earthquakes and Fractures*, Tectonophys. *5*, 35–55.

RALEIGH, B., and EVERNDEN, J. (1981), *Case for Low Deviatoric Stress in the Lithosphere. Mechanical Behavior of Crustal Rocks*, The Handin Volume, Geophysical Monograph *24*, 173–186.

SIMPSON, D. W., and NEGMATULLAEV, S. K. (1981), *Induced Seismicity at Nurek Reservoir*, Bull. Seismol. Soc. Am. *71*, 1561–1586.

SIMPSON, D. W., KEBEASY, R. M., NICHOLSON, C., MAAMOUN, M., ALBERT, R. N. H., IBRAHIM, E. M., MEGAHED, A., GHARIB, A., and HUSSAIN, A. (1987), *Aswan Telemetered Seismograph Network*, J. Geody. *7*, 195–203.

SIMPSON, D. W., GHARIB, A. A., and KEBEASY, R. M., *Induced seismicity and changes in water level at Aswan Reservoir, Egypt*. In *Induced Seismicity*, (Peter Knoll, ed.) (Central Institute for Physics of the Earth, Potsdam 1992) pp. 331–344.

TAPPOZADA, T. R., and MORRISON, P. W. (1982), *Earthquakes and Lake Levels at Oroville, California*, Calif. Geol. *35*, 115–118.

THURBER, C. H. (1983), *Earthquake Locations and Three-dimensional Crustal Structure in the Coyote Lake Area, Central California*, J. Geophys. Res. *88*, 8226–8236.

VYSKOCIL, P., TEALEB, A., KEBEASY, R. M., and MAHMOUD, S. M. (1987), *Recent Crustal Movement Studies Along the Western Bank of Lake Nasser, Egypt*, E. G. S. Proc. of the 5th Annual Meeting, Cairo, March 1987.

VYSKOCIL, P., KEBEASY, R. M., TEALEB, A., and MAHMOUD, S. M. (1989), *The Present State of Crustal Movement Studies at Kalabsha Area, Aswan, Egypt*, Internal Report, NARIAG, Egypt.

(Received March 23, 1994, revised/accepted October 12, 1994)

PAGEOPH, Vol. 145, No. 1 (1995)

0033–4553/95/010087–09$1.50 + 0.20/0

Sensitivity of a Seismically Active Reservoir to Low-amplitude Fluctuations: Observations from Lake Jocassee, South Carolina

Kusala Rajendran[1]

Abstract —Relation between water level changes and pattern of seismicity is an important consideration in studies of Reservoir Induced Seismicity (RIS). Sensitivity of the regions around Lake Jocassee to small fluctuations in the lake level is presented in this paper. The seismic source regions in the area around the lake seem to be sensitive to changes in the lake level as small as 1 to 1.5 m. Although such changes may produce stress changes of the order of only 0.1 bar, their influence on the spatial pattern of earthquakes seems to be quite perceptible. Observations of this type may help understand the threshold values of pore pressure/effective stress changes that can activate fault zones under high fluid pressure.

Key words: Induced seismicity, lake-level, seismicity pattern, pore pressure.

Introduction

Understanding the role of fluids in weakening fault zones has long been recognized as an important aspect in studying the mechanism of earthquakes. Recent studies in active fault zones provide more evidences to relate fault instability with increased fluid pressure (BYERLEE, 1990; SCHOLZ, 1990; RICE, 1992, for example). Sites of Reservoir Induced Seismicity (RIS) provide the best locations for studying the behavior of fault zones under increased pore fluid pressure. It is widely accepted that for the RIS, which results from the destabilization of critically stressed fault zones, the most important factor is the increase in pore pressure (SNOW, 1972; BELL and NUR, 1978; ROELOFFS, 1988).

Whether the earthquake is triggered by increased pore pressure due to natural processes or by a reservoir load, the important question is, "what is the critical increase in pore pressure that can really drive a fault to the point of failure?" While it may be difficult to measure the actual changes in pore pressure, it is possible to compute changes in effective stress under reasonable assumptions about crustal properties. Seismically active reservoirs are perhaps the best locations to study the threshold values of stress changes that can initiate failure.

[1] Centre For Earth Science Studies, Trivandrum 695031, India.

GRASSO et al. (1992) applied this idea at the Monteynard reservoir area to demonstrate that small changes (1 to 10 bars) in effective stress trigger earthquakes. Studies on some recent earthquakes in the San Andreas fault area suggest that stress changes as low as 0.1 bar are sufficient to induce failure (REASENBURG and SIMPSON, 1992; STEIN et al., 1992) in critically stressed zones.

Data from Lake Jocassee, South Carolina, a well documented case of RIS (TALWANI et al., 1978; GUPTA, 1992) are presented in this paper to demonstrate the influence of small changes in the lake level, on the seismicity pattern. It is observed that fluctuations of about 1–1.5 meters, which may induce stress changes of the order of 0.1 bar, can influence the pattern of seismicity in a critically stressed fault zone. Admittedly, increase in pore pressure is not the only factor that weakens a fault. A discussion on the numerous other factors that are involved in the process and their interrelations are beyond the scope of this paper. The purpose of this paper is to present an observation, which has implications on the general behavior of fault zones under increased fluid pressure.

Induced Seismicity at Jocassee Reservoir

Located in the Piedmont province in northwestern South Carolina, Lake Jocassee is a pumped storage facility (depth, 107 meters; volume $1430 \times 10^6 \text{ m}^3$). Daily changes in the lake level are generally within 0.5 m, but occasionally it may be 1 to 1.5 m. The reservoir overlies fine to medium grained Henderson augen gneiss. The rocks are heavily fractured and jointed, the prominent direction being N 10°E to N 30°E (ACKER and HATCHER, 1970). The Keowee river on which the reservoir is built also follows the same orientation.

Filling of the reservoir was completed by April 1974. Only two earthquakes were reported from within 20 km of the lake during the historic past (BOLLINGER, 1975). After the lake was filled, small tremors started occurring in the region. The first felt earthquake occurred on October 18, 1975 and the largest event ($M_L = 3.9$) was on August 25, 1979. More details are provided by TALWANI et al. (1978). Epicentral distribution of A and B quality locations suggest a general N-S trend, with an isolated cluster of events located to the southwest of the Lake (Fig. 1a). The earthquakes are generally confined to shallow depths, 77% of them occurring within 3 km (Fig. 1b).

Composite focal mechanisms indicate strike slip faulting along N-S striking nodal planes N6°E with steep easterly dips (Fig. 1a; TALWANI et al., 1978). The choice of this plane is based predominantly on the N-S trend of epicenters, better defined by earthquakes of focal depth ≥ 3 km (Fig. 2). From the epicentral trend, focal mechanisms and the orientation of fractures and joints, it is inferred that most of the seismicity at Lake Jocassee is triggered on a N-S striking fault.

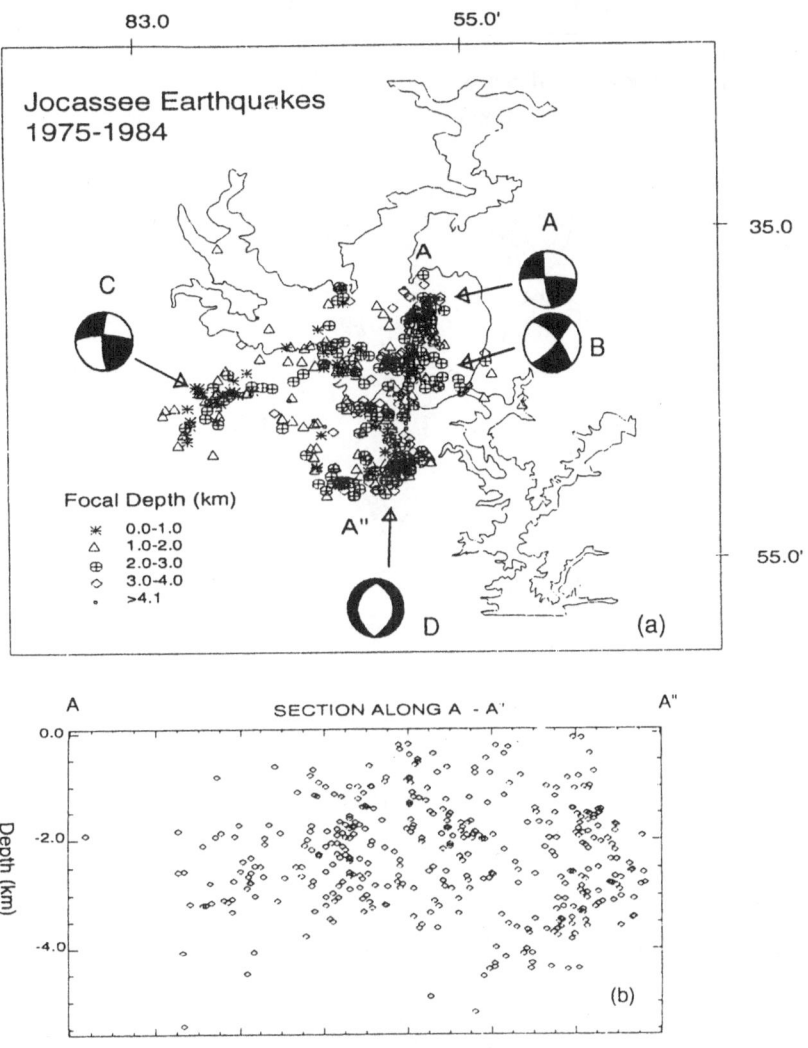

Figure 1
(a) Distribution of earthquakes at Lake Jocassee during 1975–1984. Composite focal mechanisms for groups A, B, C, and D (from TALWANI *et al.*, 1978) are also shown and (b) depth section along A–A" to which all earthquakes have been projected.

Lake Level Fluctuations and the Pattern of Seismicity

The daily changes in the water level are generally within 0.5 m, with an occasional increase to 1–1.5 m. Fluctuations over a period of six months are shown in Figure 3. This pattern is representative of the entire study period. Detailed analyses of spatial and temporal pattern of seismicity and their relation to water level fluctuations are presented by RAJENDRAN (1992). Although the frequency and

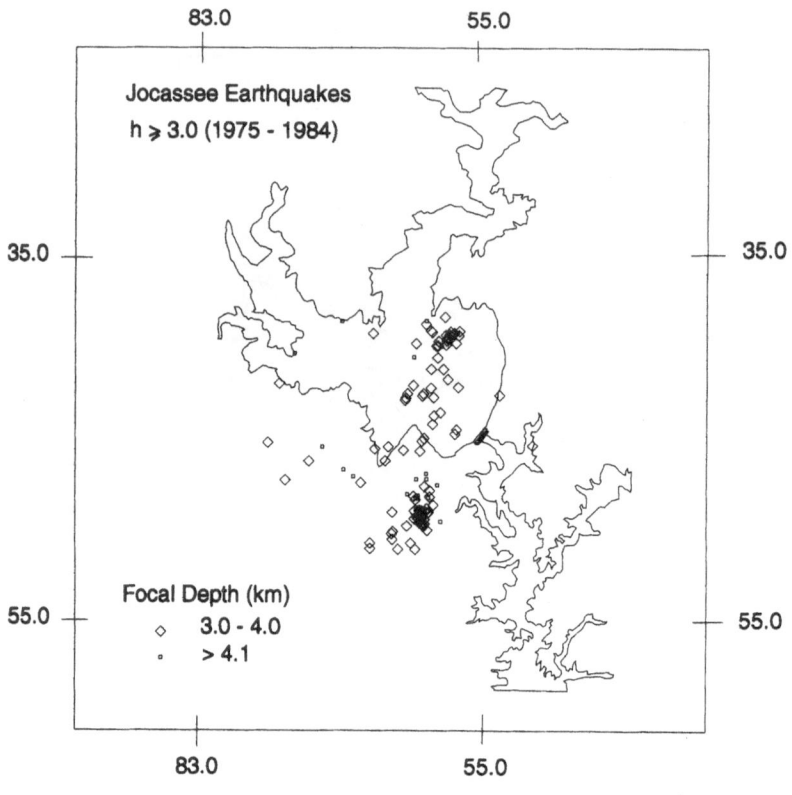

Figure 2
Distribution of earthquakes of focal depth ≥3.0 at Lake Jocassee during 1975–1984.

magnitude of earthquakes have decreased over the years, low-level seismic activity is continuing.

Earthquakes in the Lake Jocassee area are generally of $M_L < 2.0$ and they occur at random, with no systematic temporal association. Most of the seismicity at Lake Jocassee occurs below and to the south of the lake. In addition to this, there is a well-defined cluster of activity located to the southwest of the lake. Earthquakes of this cluster occur at hypocentral depths ≤ 2.0 km (Fig. 1). It may be noted that this shallow cluster occurs in the close vicinity of patches of phyllitic rocks whereas all other earthquakes occur in the Henderson augen gneiss which underlies most of the reservoir area. The earthquakes located to the south of the lake are generally deeper. Most of the earthquakes in the area with focal depth more than 4.0 km have occurred in this region (Fig. 2). These are the general spatial characteristics, but there is also a more distinct correlation between the lake level changes and the spatial distribution of earthquakes.

To identify the spatial association and its relation to the lake level changes, well located earthquakes over the period 1975–1980 were treated as two sets. The nature

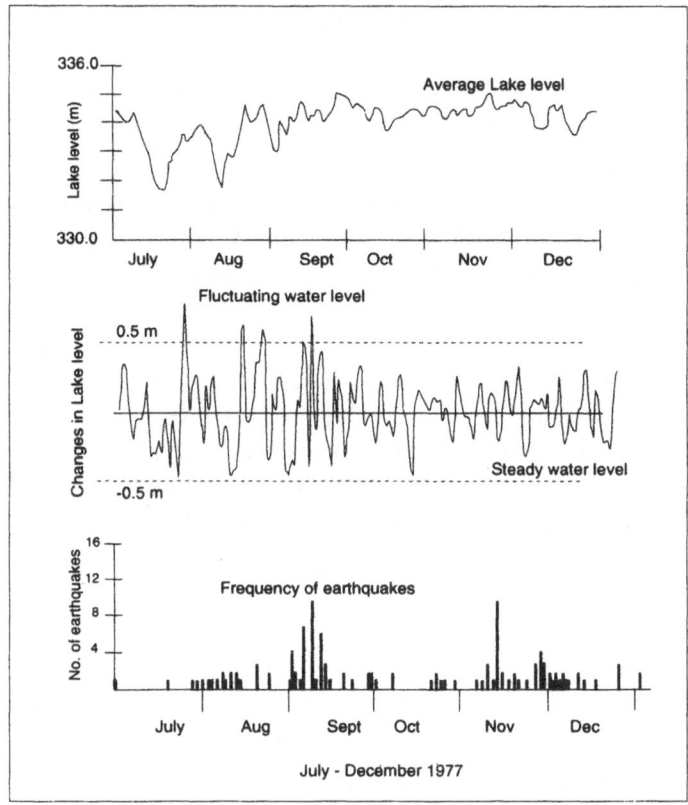

Figure 3

Average lake levels (top), daily changes in the lake level (middle) and daily frequency of earthquakes at Lake Jocassee during January–June 1976. Example of how "steady" and "fluctuating" water levels are chosen are also shown.

of variations in the lake level was considered as belonging to "periods of steady lake level" or to those of "fluctuating lake level." Periods of water level changes ≤ 0.5 m belong to the former and those with fluctuations ≥ 1.0 m to the latter. Examples of how these periods were chosen are shown in Figure 3. Epicentral distribution of earthquakes during the above periods is presented in Figures 4a and b.

An analysis of the spatial pattern of these earthquakes suggest the following. One, the earthquakes during periods of steady water level occur over a wider area. Two, a larger number of earthquakes during the steady water levels occurs at hypocentral depths ≤ 2 km (Fig. 4a). The most dramatic effect is the near disappearance of the cluster located to the southwest of the reservoir during other periods (Fig. 4b). With the exception of one event, the cluster located to the southwest of the lake occurred during periods of nominal changes in the lake level.

Discussion

The N-S alignment of epicenters together with geological evidences suggest the existence of a preexisting fault, which was weakened by the filling of the reservoir. Two-dimensional models predicting regions of failure in different fault environments (BELL and NUR 1978; ROELOFFS, 1988) suggest weakening directly below the lake on steeply dipping strike-slip faults. Since the elastic response (with zero pore pressure) in regions of vertical strike-slip faulting strengthens the regions below the lake, weakening is assumed to occur in response to pore pressure changes (ROELOFFS, 1988).

The N-S alignment of earthquakes at Lake Jocassee is compatible with the zones of weakness predicted by the 2-D models. However, the gap between the earthquakes below the reservoir and to the south of it is quite striking, particularly when the lake level fluctuates (Fig. 4b). Several of the deeper events at Lake Jocassee, including the largest event occurred in this region.

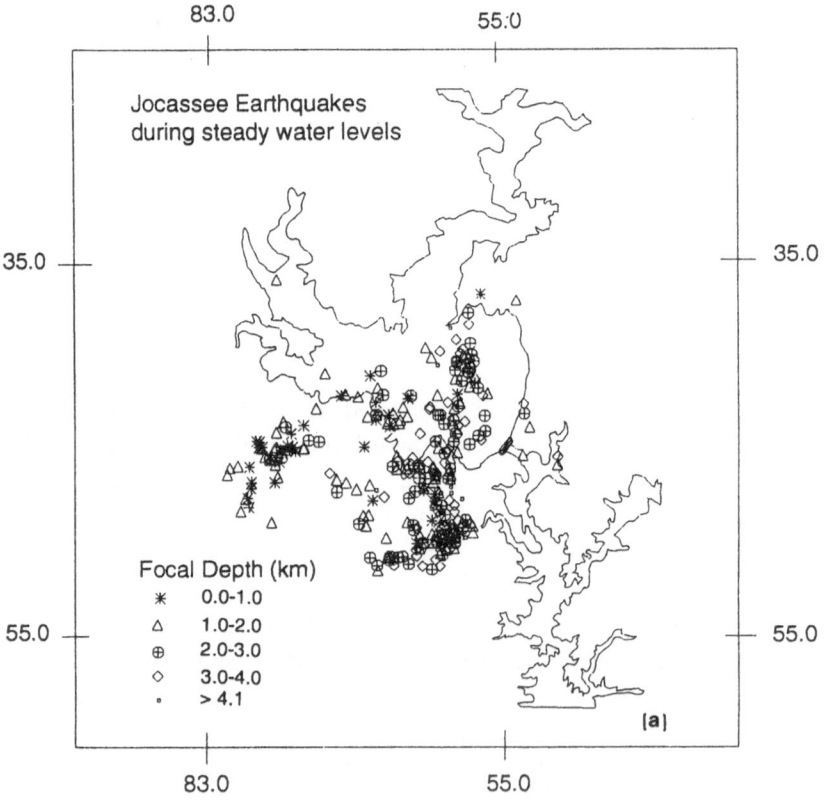

Figure 4(a)

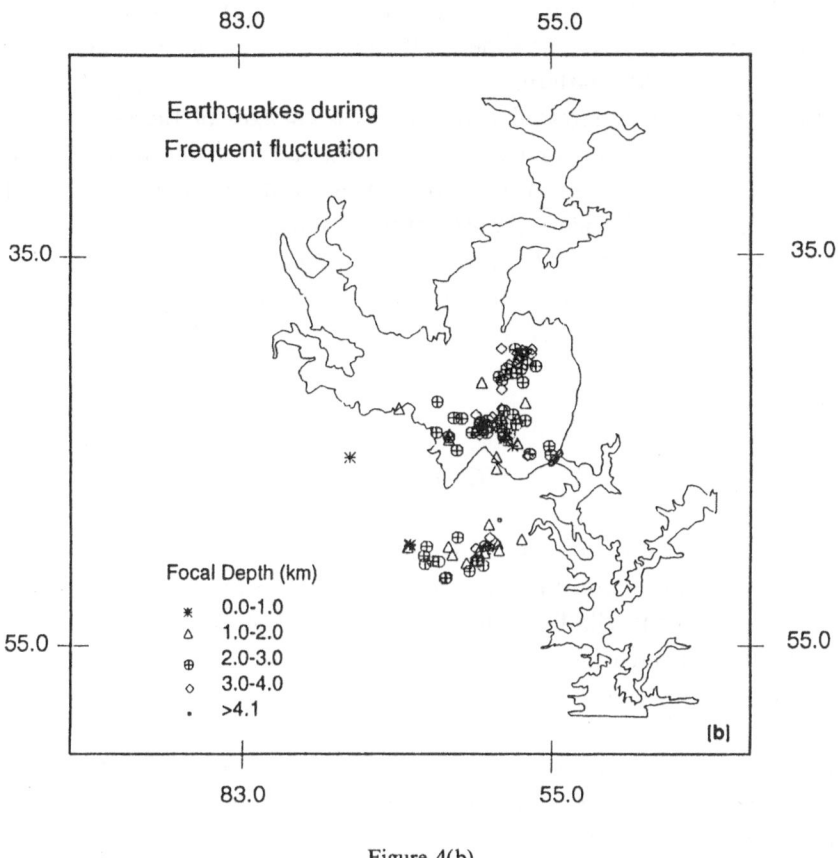

Figure 4(b)

Figure 4
(a) Distribution of earthquakes around Lake Jocassee area during steady water levels. (b) Distribution of earthquakes around Lake Jocassee area during fluctuating water levels.

TALWANI and ACREE (1984) attributed much of the seismicity at Lake Jocassee to diffusion of pore pressure. During periods of steady lake levels, pore pressure diffusion may be uninterrupted, resulting in a wider zone of weakness. Fluctuations in the lake may result in sudden changes in pore pressure. This may have maximum effect at shallow depths. With the decrease in shallow seismicity, the N-S trend becomes apparent (compare Figures 2 and 4b).

The total lack of earthquakes in the southwestern cluster during fluctuating water levels implies instantaneous transmission of stress changes to hypocentral depths. A relative decrease of seismic activity in shallow regions (≤ 2.0 km) of the N-S trending seismic zone is probably related to the same mechanism. In the absence of *in situ* measurements of crustal properties and a well constrained model for the stress changes, no quantitative explanation can be offered. However, what is important is the observation that lake level changes of the order of $1-1.5$ meter producing a stress change of the order of 0.1 bar at the surface produce perceptible changes in the seismicity pattern.

Response of critically stressed zones to stress changes of the same range have been observed at other sites of RIS as well. In a study of the initial seismicity at Monticello Reservoir, also in South Carolina, RAJENDRAN and TALWANI (1992) noted occurrence of seismicity almost within a few days of filling the lake. The computed changes in stress, at the time of their occurrence were of the same order. At Monticello, the *in situ* stress and pore pressure conditions are better known (ZOBACK and HICKMAN, 1982).

Conclusions

The spatial pattern of seismicity at Lake Jocassee suggests the following:

(i) Reservoir impoundment can weaken preexisting faults zones that are critically stressed. In a strike slip fault environment as that of Lake Jocassee, weakening occurs primarily through increase in pore pressure.

(ii) Fault zone seems to be sensitive to lake level changes of the order of $1-1.5$ m.

(iii) Changes in the effective stress may be of the order of 0.1 bar or less at the hypocentral depths in question, but they seem to be significant.

The critical change in pore pressure that may weaken a fault cannot be directly observed, but through more careful and well controlled studies, we may be able to obtain the threshold values of stress changes that may activate a particular fault. The observations presented here reiterate the importance of understanding the response of seismogenic faults to changes in fluid pressure, however small they may be. Closely monitored data from RIS sites may be useful in substantiating similar observations from other seismically active regions.

Acknowledgements

The work presented in this paper was done as a part of the author's Ph.D. programme at the University of South Carolina. The guidance and support provided by Pradeep Talwani during that period are acknowledged. The author acknowledges the study leave granted by Centre for Earth Science Studies. The idea of looking at the spatial distribution of earthquakes in relation to the lake levels was suggested by Evelyn Roeloffs. Suggestions by an anonymous reviewer have been useful.

REFERENCES

ACKER, L. L, and HATCHER, R. D. (1970), *Relationship between Structure and Topography in Northwest South Carolina*, Geol. Notes, S. C. Division of Geology *14*, 35–48.

BELL, M. L., and NUR, A. (1978), *Strength Changes due to Reservoir-induced Pore Pressure and Application to Lake Oroville*, J. Geophys. Res. *83*, 4469–4483.

BOLLINGER, G. A. (1975), *A Catalog of Southeastern United States Earthquakes 1754 through 1974*, Dept. Geol. Sci., Va, Polytech. Inst. and State Univ. Res. Div. Bull. *101*.

BYERLEE, J. (1990), *Friction, Overpressure and Fault Normal Compression*, Geophys. Res. Lett. *12*, 2109–2112.

GRASSO, J. R., GUYOTON, F., FRÉCHET, J., and GAMMOND, J. F. (1992), *Triggered Earthquakes as Stress Gauge: Implication for the Upper Crust Behaviour in the Grenbole Area, France*, Pure and Appl. Geophys. *139*, 579–605.

GUPTA, H. K. (1992), *Reservoir-induced Earthquakes*, Developments in Geotechnical Engineering *64*, Elsevier, 365 pp.

RAJENDRAN, K. (1992), *Poroelastic Models and the Mechanism of Reservoir-induced Seismicity*, Ph.D. Dissertation, University of South Carolina, 141 pp.

RAJENDRAN, K., and TALWANI, P. (1992), *The Role of Elastic, Undrained and Drained Responses in Triggering Earthquakes at Monticello Reservoir, South Carolina*, Bull. Seismol. Soc. Am. *82*, 12867–12888.

REASENBURG, P. A., and SIMPSON, R. W. (1992), *Response of Regional Seismicity to the Static Stress Changes Produced by Loma Prieta Earthquake*, Science *255*, 1687–1689.

RICE, J. R. (1992), *Fault stress states, pore pressure distributions and the weakness of the San Andreas fault. In Fault Mechanics and Transport Properties in Rocks* (Evans, B., and Wong, T. F., eds.) (Academic, San Diego 1992) pp. 475–503.

ROELOFFS, E. A. (1988), *Fault Stability Changes Induced beneath a Reservoir with Cyclic Variations in Water Level*, J. Geophys. Res. *93*, 2107–2124.

SCHOLZ, C. H., *The Mechanics of Earthquake Faulting* (Cambridge University Press 1990) 439 pp.

SNOW, D. T., *Geodynamics of seismic reservoirs. In Proc. Symp. Flow Fractured Rock* (German Soc. Soil Rock Mech., Stuttgart 1972), TS-J, 19 pp.

STEIN, R. S., KING, G., and LIN, J. (1992), *Change in Failure Stress on the Southern San Andreas Fault System Caused by the 1992 Magnitude 7.4 Landers Earthquake*, Science *258*, 1328–1332.

TALWANI, P., STEVENSON, D., SAUBER, J., RASTOGI, B. K., DREW, A., CHIANG, J., and AMICK, D. (1978), *Seismicity Studies at Lake Jocassee, Lake Keowee, and Monticello Reservoir, South Carolina (October 1977–March 1978)*, Seventh Technical Report, Contract no. 14–08–0001–14553, US Geological Survey, Reston, VA, 189 pp.

TALWANI, P., and ACREE, S. (1984), *Pore Pressure Diffusion and the Mechanism of Reservoir-induced Seismicity*, Pure and Appl. Geophys. *122*, 947–965.

ZOBACK, M., and HICKMAN, S. (1982), *In situ Study of the Physical Mechanisms Controlling Induced Seismicity at Monticello Reservoir, South Carolina*, J. Geophys. Res. *87*, 6959–6974.

(Received June 19, 1994, revised/accepted March 20, 1995)

PAGEOPH, Vol. 145, No. 1 (1995)

0033–4553/95/010097–12$1.50 + 0.20/0

The Microseismic Network of the Ridracoli Dam, North Italy: Data and Interpretations

F. G. Piccinelli,[1] M. Mucciarelli,[2] P. Federici,[2] and D. Albarello[3]

Abstract—The Ridracoli Dam has been operating since 1981. Around the reservoir ISMES installed and operated for 10 years a seismic network, now reduced to a 3-D station. Earthquakes were recorded with completeness from magnitude 0.8 onwards. In the same period, all the parameters relevant to the dam and the environment were measured. This provided a complete data base for RIS studies, unique in its kind in Italy. The main findings of the analyses performed are the following:

1) The filling of the reservoir has not influenced the seismicity of the area for most significant events ($M_L > 3.5$).

2) Lesser seismicity around the reservoir seems to be correlated with water level in the reservoir, but also shows to be dependent on regional seismicity.

3) b value shows a slight increase with time. This may indicate an increase in rock fracturing, which is known to precede the disappearing of Type II RIS.

Key words: Microseismic, reservoir-induced seismicity, nonparametric correlation analysis, b values.

Introduction

In the past, several cases of reservoir-induced seismicity were observed. A list of the best known cases which occurred all over the world can be found in the studies done by Gupta (1984), Simpson *et al.* (1988) and, for Italy, by Caloi (1970). Following Simpson *et al.* (1988) it is possible to identify two kinds of seismicity (sometimes observable in the same basin) induced by reservoir activities.

The first group (Type 1) concerns those earthquakes that occur in the first phase of the filling and soon after the filling of the reservoir, following a rapid change in water level (undrained response). In this typology there are:

a) particularly intense earthquakes due to a water lubrication effect on buried structures already under tectonic load that, in a dry regime, did not overcome the threshold of frictional resistance needed to cause an earthquake;

---

[1] Consorzio Acque per le province di Forlì e Ravenna, Piazza del Lavoro 35, 47100 Forlì, Italy.

[2] Ismes S.p.A., Divisione Geofisica, viale G. Cesare 29, 24100 Bergamo, Italy.

[3] Università di Siena, Dipart. di Scienze della Terra, Istituto di Geofisica, Via Banchi di Sotto 55, 53100 Siena, Italy.

b) seismic activity of the "swarm" type, referring to the basin area, often widespread on a certain volume of rocks rather than on a well-defined fault surface, with a low magnitude and a shallow hypocentral depth (< 10 km). This activity, strictly correlated with changes in water level of the reservoir, is due to the increase in elastic stress, as a response to the superficial load. An increase of pore pressure due to compaction of pore space is associated: increase of stress (and thus seismicity) follow the load variation with slight delay, or no delay at all.

The second group (Type 2) concerns those earthquakes that occur after a certain period of time from the beginning of the reservoir impounding (delayed or drained response): the magnitude is variable, often high; the hypocenters, deeper than in Type 1, can be extended over an area wider than the basin borders (10 km or more); sometimes there is evidence of a correlation with fault zones crossing the basin. The "delayed response" can arise from the diffusion of the pore pressure and the water flux outside the basin coupled with the elastic load. These phenomena can experience a considerable delay from the moment in which the surface load starts.

In this theoretical framework, possible seismic effects have been analysed for the Ridracoli reservoir and dam (Northern Italy). Main characteristics are a dam height of 103 m, a maximum water height of 80 m and a reservoir storage capacity of 33 million m^3. The dam is located within one of the most active seismogenic zones in Northern Italy (Forlivese province) affected in the past by destructive earthquakes with site intensities reaching IX–X of Mercalli, Cancani, Sieberg (MCS) scale (BOCCALETTI *et al.*, 1985).

Due to this situation, microseismicity has been surveyed in the reservoir area during approximately 9 years in order to detect possible geodynamical perturbation induced by dam activities. Since at the early stages of the dam operation's life no Type 1 induced seismicity was observed, in this study we concentrated our research on Type 2 events.

Seismicity which occurred from 1981 to 1989 in an area 15 km from the reservoir is analysed. Furthermore, in order to investigate the possible interrelations between reservoir activity and seismicity, a smaller area within 5 km from the dam has been considered. Seismicity of this subset of data is described in terms of occurrence in time and released seismic energy (*b* parameter of the Gutenberg and Richter law). A nonparametric analysis was conducted to state the significance of the effect of changes in water level on seismicity.

Geological and Seismotectonic Framework of the Area

Ridracoli basin is on the Adriatic side of the Monte Falterona ridge, in the area of the Marnoso-Arenacea formation outcrop in romagnan facies. It is composed of detritic turbiditic grounds relayered in the Langhian-Tortonian period. The formation consists of an alternation of four fundamental lithologic types: sandstones, silts, marls and clays. Sometimes, these are associated with some conglomerates, calcarenites and marly limestones. Quartz-feldspathic sandstones prevail and the formation has a tabular geometry in benches from 30 cm to some meters thick, with a very large horizontal continuity. In the area that directly concerns the foundation of the dam, benches have a monoclinalic trend dipping about 30° (REBAUDI, 1978). The thickness of the entire formation is about 5300 m. Its lower limit is tectonic while the upper is stratigraphic and is constituted by the base of the Gessoso-Solfifera Messinian Formation (RICCI LUCCHI, 1967).

In the North Apennines framework, the area of the romagnan authoctonous, from the point of view of its structural evolution and crustal structure, is part of the "External belt" or "Principal" (BOCCALETTI et al., 1985). It is characterized by crustal thickness of about 35 km as suggested by gravimetric, aeromagnetic and seismic investigations. Neotectonic analysis suggests recent upliftment activity.

In the surroundings of Ridracoli dam, the Marnoso-Arenacea is characterized by dislocation lines trending NW-SE (SIGNORINI, 1940). Two of these tectonic features (Fig. 1) runs about 3 km southward of the dam, and about 2 km northward, respectively. These structures are reported as reverse faults (with a southwest dipping plane) on the Geological Map of Italy (MERLA and BORTOLOTTI, 1969). Other Apennine trending structures have been detected by photo interpretation just north of S. Sofia (GELMINI, 1966). In the Neotectonic Map of the Northern Apennines (BARTOLINI et al., 1982) a fault running near S. Sofia and parallel to the Bidente River with an anti-Apenninic trend is found. It is dated at 0.7 Ma. CASTELLARIN et al. (1985) considered the line formed by the Bidente River-Ronco River as one of the greatest transversal lines of the Northern Apennines. It is presumed to have a left transcurrent character and a vertical activity, mainly in the Late Messinian and in the Late Pleistocene. The same local structure and three other minor ones were studied by MARABINI et al. (1985). They run NW parallel to the first one and are interpreted as subvertical faults.

The seismicity of the zone seems to be connected with the structural discontinuity of the Marnoso-Arenacea formation and the Mesozoic basement, from which it seems to be partially detached. Based on the historical data, the Forlivese province (more specifically the area between the Bidente and the Lamone Rivers) is in the Emilia-Romagna region, the area in which the earthquakes reached the highest intensity ($I_{MCS} = IX-X$). According to PATACCA and SCANDONE (1986) the earthquakes of Predappio, 1661, ($I_{MCS} = IX-X$), S. Sofia 1768, ($I_{MCS} = IX-X$) and Forlí, 1781, ($I_{MCS} = IX$) have been generated by two crustal seismogenic structures called the Ferrarese-Romagnan Arch and the Forlivese Track. Along the Ferrarese-Romagnan Arch, earthquakes with thrust mechanisms are expected, with fault directions NW-SE. In the Forlivese Track, earthquakes having right lateral strike-slip faulting on N 20°E to N-S fault planes are expected. Faults related to two structures are present in the examined zone (Fig. 1).

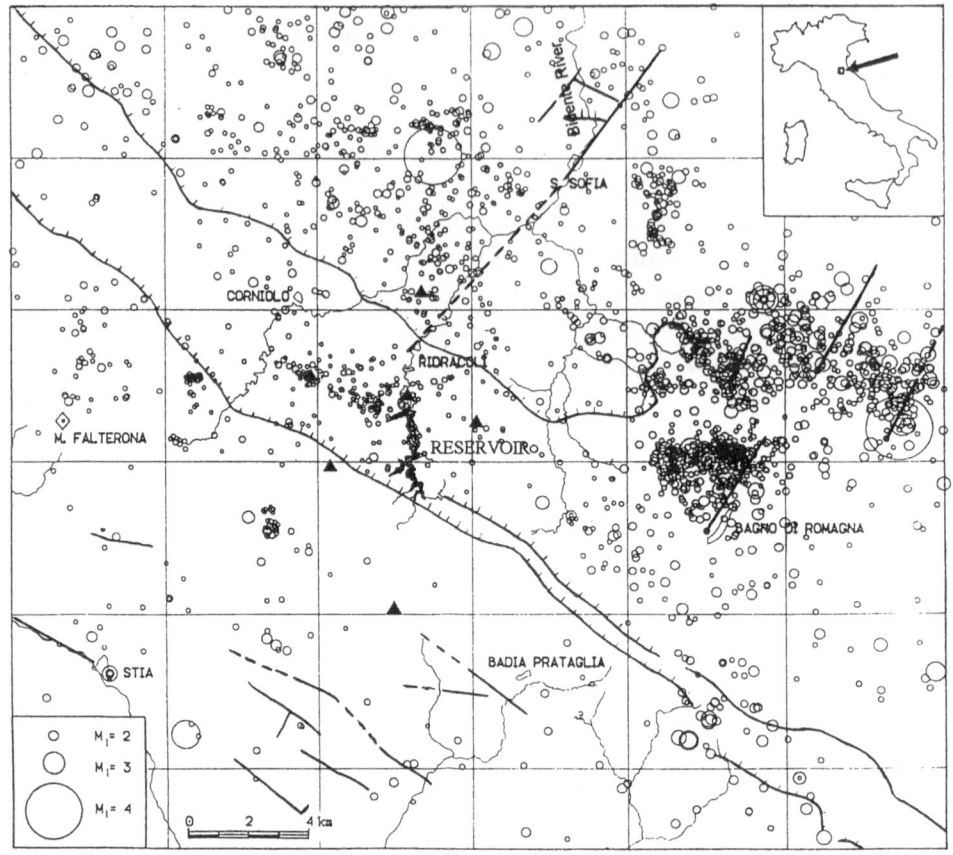

Figure 1

Seismicity recorded by the microseismic network from 1981 to 1989. Dimensions of circles proportional to focal volumes, triangles indicate seismic stations. Faults are represented by solid lines (dashed when uncertain). Main tectonic features running NW-SE are reverse faults dipping SW, hatches indicate downthrown side. Structures running SW-NE near S. Sofia are presumed to be subvertical and have left transcurrent characters.

The Network and the Surveyed Microseismicity from 1981 to 1989

The Ridracoli microseismic network operated from November 1981 until December 1989. It was comprised of a group of six stations, one of which was equipped with a 3-component seismometer, while five stations had vertical component seismometers. The 3-D station was near the dam and the vertical stations surrounded the basin at variable distances from about 3 to 7 km (Fig. 1).

The seismic signals were transmitted by analog telemetry to a recording centre. The signals were digitized and recorded with a sampling rate of 200 samples/sec.

Earthquakes were located using the Hypoellipse program (LAHR, 1984) and a velocity model with plane and parallel layers inferred from direct measurements, for shallower layers, and from a synthesis of reflection seismic surveys (ISMES, 1989) for deeper structures. Either duration or local magnitudes (M_L and M_D, respectively) have been determined for the detected earthquakes.

In more than eight years, the microseismic network recorded more than 2600 local events in a range of about 15 km from the dam. A global representation of the observed seismicity is shown in Figure 1.

Hypocenters detected by the seismic network were mainly shallow and concentrated in the upper crust (about 85% of the events less than 10 km deep) with the highest frequencies (especially in the area near the basin) at a depth of 4–5 km. The regional seismicity seemed to be mainly concentrated east-northeast of the dam. The zone of greatest activity was located north of Bagno di Romagna, where the seismic sequences of greatest intensity occurred (about 250 events with $M_L \leq 4.1$ in November and December 1985, about 230 events of $M_L \leq 3.3$ in August 1986, about 110 events of $M_L \leq 2.3$ in February 1987 and over 300 events of $M_L \leq 3.1$ in July 1989). The seismicity in this area corresponds to a set of anti-Apenninic trending faults (Fig. 1).

In the area near the dam, the microseismicity was mainly concentrated in two zones. The first one is part of a general trend parallel to the NW-SE compressive structure running among Corniolo, Ridracoli and Bagno di Romagna. It is characterized by hypocentral depths ranging between 5 and 10 km. The second one is located between Corniolo and Ridracoli, within a belt trending ESE-WNW. Hypocentral depths in this second zone range between 3 and 5 km. In the first zone, the two events of the highest magnitude ($M_L = 4.0$ and $M_L = 2.8$) occurred at the epicentral distance of 9 and 6 km from the dam respectively, and at 14 and 7 km of depth. The two events seem to be isolated unlike the Bagno di Romagna zone which is characterized by seismic swarms. In the second zone, the event with the highest magnitude ($M_D = 2.5$) occurred a few hundred meters from the dam at about 3 km of hypocentral depth.

During the time interval considered, seismicity showed significant fluctuations both in terms of occurrence rate and energy release. However, this pattern cannot be simply considered as representative of true variations of seismic process. In fact, following the approach proposed by MULARGIA et al. (1987), the observed seismicity ("apparent seismicity") is the result of a complex convolution of the "true" seismicity and a "completeness" function which represents deterministic man-induced effects (variations in network configuration, trigger thresholds, etc.). Thus, for the purpose of assessing the significance of the observed fluctuations, a preliminary analysis must be performed in order to select a "complete" data subset. For this purpose, only earthquakes which occurred within 5 km from the dam (i.e., the area included within the network) have been considered.

As far as the magnitude threshold is concerned, a completeness analysis has been conducted following the approach proposed by BÅTH (1983). In this approach, it is assumed that the "true" seismicity follows the well-known Gutenberg and Richter law (see, e.g., LOMNITZ, 1974): those deviations from the law of the empirical magnitude-frequency relationships are interpreted in terms of incompleteness of the data set. The magnitude range satisfying the Gutenberg and Richter law selects the "complete" data set. Following this approach, the magnitude threshold has been fixed at 0.8 and only earthquakes with magnitudes above this value have been considered in the following analysis. This complete sample resulted in 80 earthquakes during October 1982–1989.

Even a superficial investigation of seismic rate in the complete data set (Fig. 2) points out that July and August 1985 (a few months after the maximum water level was reached) represent a turning point in the microseismicity pattern. Futhermore, no induced seismicity due to undrained response is observed. Seismicity rate variations and also the occurrence of the highest magnitude earthquake seem to be delayed with respect to water level changes.

One can also remark that the seismicity cannot be easily interpreted within the framework of a simple Poissonian process with constant rate. Particularly evident is the existence of two periods without any seismicity: from March to November 1983 and from August 1984 to May 1985. Furthermore, it is possible to recognize the presence of periods with a different activity rate, and in particular the period starting from the second half of 1985, when the rate reached a value significantly

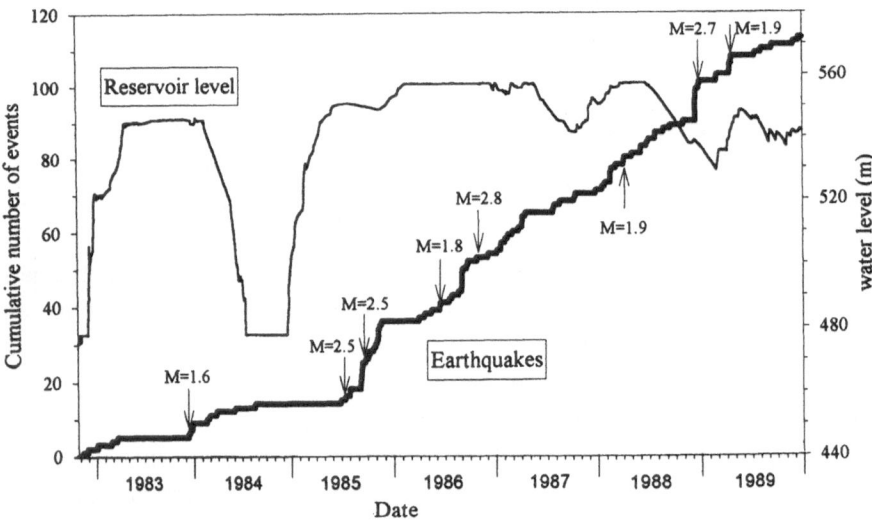

Figure 2
Cumulative number of events within 5 km from the Ridracoli Dam of magnitude $M_L \geq 0.8$ (thick line, scale on the left) and water level (thin line, scale on the right). Events of the highest magnitude are indicated.

higher than the one recorded in the 1982–1985 interval (0.050 events per day versus 0.015 events per day of the previous period).

Other features of the observed seismicity suggest the presence of a time-varying geodynamic situation. In fact, during the period under examination, seismicity monitored by the local seismic network manifested different partitions of the released seismic energy. In order to show this phenomenon, the average magnitude of earthquakes within discrete time intervals has been computed. This parameter is found to be related to parameter b of the Gutenberg and Richter law (e.g., LOMNITZ, 1974) which, in turn, is representative of rock fracturing and the stress condition of the area under study (MOGI, 1962; SHOLTZ, 1968; URBANCIC et al., 1992). Figure 3 displays the time pattern of b values and suggests, in spite of the experimental errors associated with this kind of estimate, the presence of a systematically increasing pattern.

Seismicity and Reservoir Level

In order to check the statistical significance of possible interrelations between seismicity rate fluctuations and the water level of the dam, a nonparametric correlation analysis (KENDALL, 1938) has been carried out. Applications of this approach to the analysis of seismic series are reported by MUCCIARELLI and ALBARELLO (1991) and ALBARELLO et al. (1989). The adopted statistical approach allows recognition of monotonic deterministic interrelations and the assessment of their significance level irrespective of the parent frequency distributions of the series considered. Furthermore, the results obtained following this approach are

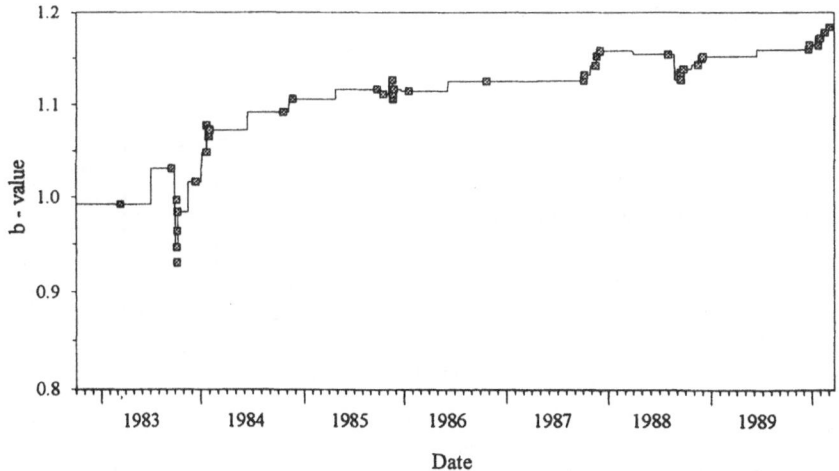

Figure 3
Temporal variation in b values. It is computed by means of a maximum likelihood estimate, starting from the average magnitude of earthquakes within discrete time intervals.

insensitive to monotonic transformations of the involved variables. This means that, for the seismicity, it makes little or no difference to use the intensity, the magnitude, the energy or the strain release because all these quantities are connected with one another by a monotonous functional relationship. The same is true for the volume and level of the reservoir.

In the following, the presence of possible interrelations between dam activity and seismicity has been analyzed, taking into account the daily water level in the reservoir and seismic rates in the complete data set.

The analysis of the nonparametric cross correlation pointed out that the impounding level of the reservoir is correlated to the seismicity in a positive way (Fig. 4). In particular, the maximum level of interrelation corresponds to a relative delay between the two series ranging from two to five months (solid line in Fig. 4, and maximum of correlation coefficient τ at negative values of delay). This correlation results to be statistically significant over a 0.01 threshold level (KENDALL, 1955). This means that the maximum values of the reservoir level tend to precede the maximum values of the seismicity rate (the same applies to the minimum values).

This result, however, might be biased by the particular structure of the data set. In fact, the time series of water levels presents a first period (until the first half of 1985) in which two complete charge/discharge cycles have performed (see Fig. 2) with an average seasonal fluctuation in the lake level of about 70 m, while in the second period the level was relatively stable with an average seasonal fluctuation of

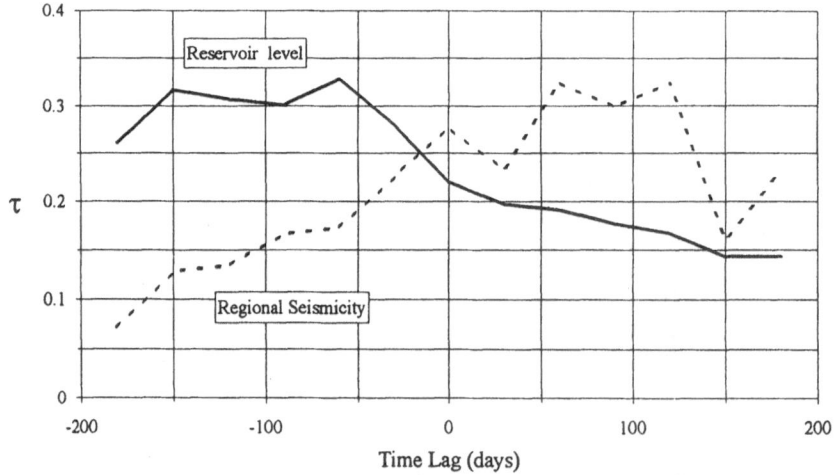

Figure 4
Cross correlations between microseismic activity within a 5 km epicentral distance from the dam and reservoir level (solid line) and between microseismic activity and regional seismic activity (the entire set of earthquakes recorded by the local seismic network) within about a 30–35 km epicentral distance from the dam (dashed line). The maximum value of cross-correlation coefficient occurs at a 60-day time lag. See text for further explanation.

about 25 m. This variation roughly corresponds to the seismic rate turning point as mentioned above.

The presence of this trend common to the two series should be taken into account when the cross-correlation analysis is performed (see, e.g., DAVIS, 1986). In order to do this, the cross-correlation analysis has been performed again, dividing the studied period into two parts: before and after the turning point in the water level and seismicity rate patterns in mid-1985. This analysis revealed that the correlation level between reservoir water level and seismicity is significant as concerns the first period, while it becomes worse in the second period. So as to provide a deeper insight of the geodynamical meaning of these findings, the relationship has been analyzed between local seismic activity (within 5 km from the dam) and the entire set of earthquakes detected by the seismic network. A significant correlation (over the threshold level of 0.01) has been found (dashed line in Fig. 4). In particular, the local seismicity pattern seems to coincide or to precede in time the regional seismicity, suggesting a relative geodynamical dependence of seismicity at the local and regional scale. Thus, the increase of seismic rates observed in the restricted area after August 1985 seems to coincide with a general increase in seismicity at a wider scale.

Some further indications regarding the possible dynamical dependence of water level and seismicity can be obtained by the analysis of the seismicity reported in a wider area (from 10 up to 50 km from the dam) by the Italian National Catalogue (POSTPISCHL, 1985), updated until 1990 with *Istituto Nazionale di Geofisica* (ING) bulletins. Seismic events with magnitudes greater or equal to 3.5 (or $I_{MCS} \geq V$) which occurred within 10, 25 and 50 km from the dam have been considered before and after the building of the dam (Fig. 5). The occurrence of earthquakes produced several significant fluctuations in the period of "completeness" (1880–1988). However, in the last 10 years, roughly corresponding to the period of dam activity, the seismic rate was slightly lower than the average computed over the entire data set for any distance considered.

Conclusions

The seismicity which occurred in the neighborhood of the Ridracoli dam (Northern Italy), monitored from 1981 until 1989 by a local microseismic network, has been analyzed in connection with the reservoir level.

The general seismicity pattern exhibits a NW-SE striking parallelism with the south-westward dipping reverse faults. The microseismicity nearest the basin seemed to be mainly concentrated along an alignment starting slightly north of the dam, with a length of about 6 km, and a rather significant Apenninic direction. No geological evidence of this alignment, however, is reported in the literature, but two parallel compressive structures are present at distances of about 2 and 3 km.

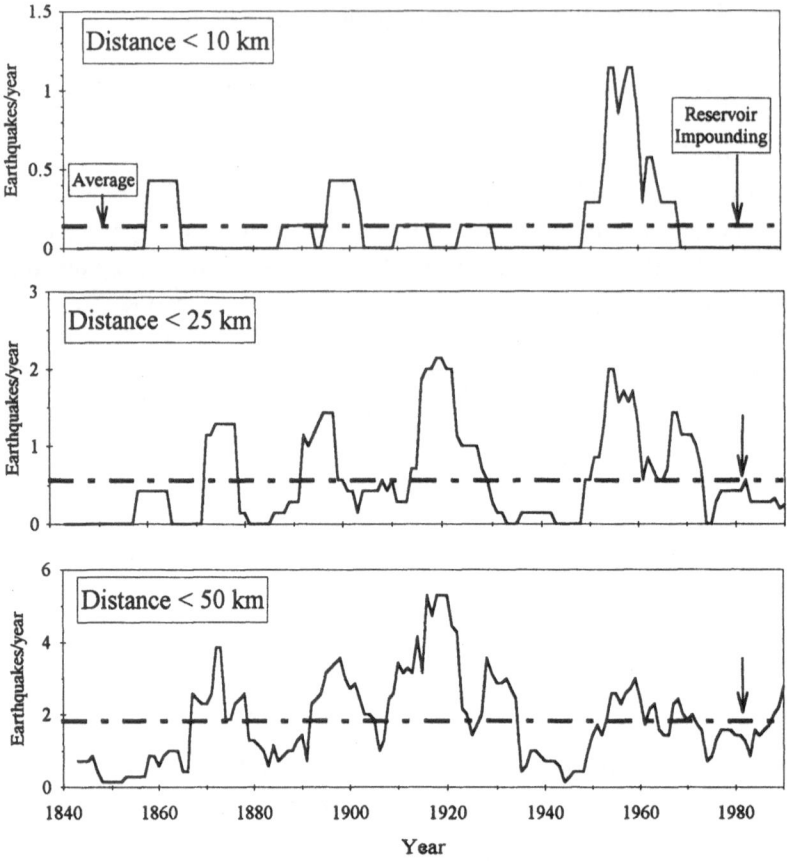

Figure 5

Occurrence of earthquakes within 10, 25 and 50 km from the dam of magnitude $M \geq 3.5$ or $I_{MCS} \geq V$ (solid lines). Data from the Italian National Catalogue (POSTPISCHL *et al.*, 1985) updated until 1988 with ING bulletins. Dashed line indicates the average occurrence rate.

Seismicity at the regional scale (up to 50 km from the dam), which occurred during the decade 1980–1990 corresponding to the dam activity, has been slightly lower than the average recorded in the last century. This suggests that the Ridracoli reservoir did not yield significant variations in the seismicity of average to high energy (M_L greater or equal to 3.5) of the Tuscan-Emilian Apennine.

The correlation analysis between the reservoir level and the seismic rate concerning earthquakes with $0.8 \leq M \leq 2.8$ recorded in a range of 5 km from the basin, reveals the presence of a statistically significant interrelation. It indicates that the maximum values of the reservoir water level precede by some months, the microseismicity near Ridracoli. The correlation between microseismicity and reservoir level is better if we consider the first two impoundings of the basin from 1982 to 1985, while it becomes worse in the following period when the reservoir water level was more or less constant and there was a resumption of the seismic activity in the entire area. This evidence suggests that from 1982 to 1985, during the period of relative quiescence of the whole North-Apennine area, the microseismic activity could be realistically attributed to the variations of the reservoir level. In the following period (after 1985), the coincidence between the presence of a generalized seismic activity and the constant maximum impounding level makes it impossible to determine which part of the microseismicity can be attributed to the effect of the reservoir.

A slight increase of the b value has been observed, which may correspond to an intensification of the rock fracturing; this type of finding could be interpreted as a signal of progressive inhibition of reservoir-induced seismicity: in fact, due to the increasing fracturing of the rocks in the reservoir area, these rocks become unable to store enough deformation elastic energy to cause a significant increase of seismic rate.

REFERENCES

ALBARELLO, D., MUCCIARELLI, M., and MANTOVANI, E. (1989), *Use of Nonparametric Correlation Test for the Study of Seismic Interrelations*, Geophys. J. *96*, 185–188.

BARTOLINI, C., BERNINI, M., CARLONI, G. C., COSTANTINI, A., FEDERICI, P. R., GASPERI, G., LAZZAROTTO, A., MARCHETTI, G., MAZZANTI, R., PAPANI, G., PRANZINI, G., RAU, A., SANDRELLI, F., VERCESI, P. L., CASTALDINI, D., and FRANCAVILLA, F. (1982), *Carta Neotettonica dell'Appennino Settentrionale*, Note illustrative Bollettino Società Geologica Italiana *101*, 523–549.

BÅTH, M. (1983), *Earthquake Data Analysis: An Example from Sweden*, Earth Sci. Rev. *190*, 181–303.

BOCCALETTI, M., COLI, M., EVA, C., FERRARI, G., GIGLIA, G., LAZZAROTTO, A., MERLANTI, F., NICOLICH, R., PAPANI, G., and POSTPISCHL, D. (1985), *Considerations on Seismotectonic of the Northern Apennines*, Tectonophys. *117*, 7–38.

CALOI, P. (1970), *How Nature Reacts to Human Intervention. Responsibilities of Those who Cause and Those who Interpret such Reactions*, Annali di Geofisica *23*, 283–305.

CASTELLARIN, A., EVA, C., GIGLIA, G., and VAI, G. B. (1985), *Analisi strutturale del fronte appenninico padano*, Giornale di Geologia *3*, 47/1–2, 47–76.

DAVIS, J. C., *Statistics and Data Analysis in Geology* (Wiley and Sons 1986) 646 pp.

GELMINI, R. (1966), *Studio Fotogeologico dell'Appennino Settentrionale tra il Valdarno e la Romagna*, Bolletino Società Geologica Italiana *84*, 167–212.

GUPTA, H. K. (1984), *The present status of reservoir-induced seismicity investigation with special emphasis on Koyna earthquakes*. In *Quantification of Earthquakes* (eds. Duda, S. J. and Vanek, J.) Tectonophys. *118*, 257–279.

ISMES, Internal Report (1989), Diga di Ridracoli. Rete di controllo microsismico del bacino. Analisi della microsismicità rilevata nel periodo 1981–1988, Rapporto per conto Cons. Acque Prov. Forli e Ravenna.

KENDALL, M. G. (1938), *A New Measure of Rank Correlation*, Biometrika *30*, 81–93.

KENDALL, M. G., *Rank Correlation Methods* (Griffin and Co., London 1955).

LAHR, J. C. (1984), *Hypoellipse/VAX: A Computer Program for Determining Local Earthquake Hypocentral Parameters, Magnitude and First Motion Pattern*, U.S. Geol. Survey Open File Report, 84–519.

LOMNITZ, C. (1974), *Global Tectonics and Earthquake Risk*, Developments in Geotectonics *3*, 320 pp.

MARABINI, S., BALDI, P., BENINI, A., MULARGIA, F., VALENSISE, G., and VAI, G. B. (1985), *Strutture Tettoniche da Monitorare nei Dintorni di S. Sofia (Appennino Forlivese)*, Atti del 4° Convegno CNR–GNGTS, Roma.

MERLA, G., and BORTOLOTTI, V. (1969), *Note Illustrative della Carta Geologica d'Italia*. F. 107 M. Falterona, Ercolano.

MOGI, K. (1962), *Study of Elastic Shocks Caused by the Fracture of Heterogeneous Materials and its Relation to Earthquake Phenomena*, Bull. Earthq. Res. Inst. *40*, 125–173.

MUCCIARELLI, M., and ALBARELLO, D. (1991), *The Use of Historical Data in Earthquake Prediction: An Example from Water Level Variation and Seismicity*, Tectonophys. *193*, 247–251.

MULARGIA, F., GASPERINI, P., and TINTI, S. (1987), *A Procedure to Identify Objectively Active Seismotectonic Structures*, Boll. Geofis. Teor. Appl., *XXIX, 114*, 147–164.

PATACCA, E., and SCANDONE, P. (1986), *Struttura Geologica dell'Appennino Emiliano-Romagnolo: Ipotesi Sismotettoniche*. Regione Emilia-Romagna, Convegno regionale die Cartografia, Atti del Seminario, Bologna, 22–23 febbraio 1985, pp. 102–118.

POSTPISCHL, D. (1985), Catalogo dei Tenemoti Italiani dall'Anno 1000 al 1980, C.N.R., quaderni della ricerce scientifica, 114, 2B.

REBAUDI, (1978), *Cenni Informativi Circa le Indagini e le Prove Esperite per la Ricerca dell Proprietà Fisiche e Meccaniche della Roccia di Fondazione della Diga di Ridracoli*, Atti del 13° Convegno Naz. di Geotecnica, Merano, 2, 75–87.

RICCI LUCCHI, F., *Formazione Marnoso-Arenacea Romagnola*. IV Congresso Neogene Mediterraneo, Guida alle escursioni (ed. Selli, R.) (Bologna 1967) pp. 111–120.

SCHOLTZ, C. H. (1968), *The Frequency-magnitude Relation of Microfracturing and its Relation with Earthquakes*, Bull. Seismol. Soc. Am. *58*, 399–441.

SIGNORINI, R. (1940), *Sulla tettonica dell'Appennino Romagnolo*. Atti Regia Accademia d'Italia, Rend. Ser. 7, 1, Roma.

SIMPSON, D. W., LEITH, W. S., and SCHOLTZ, C.H. (1988), *Two Types of Reservoir-induced Seismicity*, Bull. Seismol. Soc. Am. *78*, 2025–2040.

URBANCIC, T. I., TRIFU, C. I., LONG, J. M., and YOUNG, R. P. (1992), *Space-time Correlations of b Values with Stress Release*, Pure and Appl. Geophys. *3/4*, 449–462.

(Received April 22, 1994, revised/accepted November 8, 1994)

PAGEOPH, Vol. 145, No. 1 (1995)

0033-4553/95/010109-14$1.50 + 0.20/0
© 1995 Birkhäuser Verlag, Basel

Investigations of June 7, 1988 Earthquake of Magnitude 4.5 near Idukki Dam in Southern India

B. K. RASTOGI,[1] R. K. CHADHA,[1] and C. S. P. SARMA[1]

Abstract — A mildly damaging earthquake of magnitude 4.5 and intensity VI occurred 20 km east of the Idukki reservoir, Kerala in southern India. With a network of 5 seismic stations, the aftershocks which continued for 3 1/2 months were monitored. The hypocentral parameters, b value, M_1/M_0 ratio indicate that this earthquake sequence does not qualify to be categorized as induced. The trend of the aftershocks, composite fault plane solution and local tectonics point towards reactivation of a NW-SE fault along the Kallar river. The existence of such a fault is also supported by gravity studies.

Key words: South India, earthquake sequence.

Introduction

A mild damaging earthquake of magnitude 4.5 and intensity VI on the MM scale occurred on June 7, 1988 some 20 km east of the mammoth Idukki Dam in Kerala. The earthquake was felt in the Idukki and Kottayam districts of Kerala and the Madurai district of Tamilnadu in an area of about 50 km radius. This earthquake and recurring aftershocks caused panic because of strong shaking, damage to several houses and nearness to the Idukki Dam. About 5000 dwellings suffered slight to moderate damage. The damage was more concentrated in the towns of Nedumkandam and Kallar. We carried out intensity surveys and monitored seismicity from June 21, 1988 until the end of the sequence with a 5 seismic stations network. The study of this sequence produced important results for the understanding of seismicity of the stable continental region of India. These results have assumed greater importance due to the increased level of seismicity in the region and especially after the occurrence of a destructive earthquake of $m_b = 6.4$ in the Latur-Osmanabad area in 1993.

---

[1] National Geophysical Research Institute, Uppal Road, Hyderabad-500007, India.

Particulars of the Idukki and Other Nearby Dams

Idukki Dam (9°50'N, 76°58.5'E) is about 120 km east of Cochin (Ernakulam) across the Periyar River. The height of the dam is 169 m with a capacity of 2 km³. Impounding began in 1974 and the reservoir was filled to capacity in 1981. There are several large and small dams nearby and others are being constructed. Notable among them are: Kakki about 50 km south of Idukki, with a height of 110 m and capacity 0.46 km³, constructed in 1966. Idamalayar approximately 50 km NW of Idukki, with a height of 94 m and capacity 1.08³, was constructed recently. Poringalkuthu and Sholayar, constructed in 1965 and 1967, respectively and situated at a distance of about 60 km NW of Idukki, are of medium size. A large dam is to be constructed 12 km NNW of Idukki at Lower Periyar.

Seismicity of Kerala

For the majority of the local population the earthquake sequence in June 1988 was their first experience. However, scrutiny of a catalogue revealed that 15 small to moderate earthquakes have been experienced in Kerala (GUHA *et al.*, 1970; PADALE and DAS, 1988). These have occurred on the Malabar coast in Travancore, Trivandrum, Calicut and Tellichery-Cannanore. Some small shocks were reported recently at different sites at Kottuli near Calicut in north Kerala, at Boothathankettu (Idamalayar), at Pathanamthitta, at a site between Calicut and Tirur nea Palode north of Trivandrum. In another notable case, 257 shocks were reported to be felt in the Mangalam Dam area (10°31'N, 76°32'E) from March 1968 to January 1971. Out of these, seven tremors were strongly felt (PADALE and DAS, 1988). The Idukki network of seismic stations has operated since 1971 and recorded a large number of small shocks. Only a few of these were felt locally. Thus, various regions of Kerala are not free of seismic activity. But it is also true that none of these tremors (except for the one near Palode) are reported to have caused any damage. Hence, the tremor on June 7, 1988 is the only earthquake in Kerala to have produced serious damage. In the Seismic Zoning Map of India, prepared by the Indian Standards Institution (ISI), Kerala lies in zone III, where moderate earthquakes are expected.

Geology

The rock types in the area are mainly Charnockites-Khondalites and gneisses. The NW-SE trending Periyar River is a major lineament in the area (Fig. 1). The present seismic activity occurred along the NW-SE flowing Kallar River which is a continuation of the Periyar trend in this area. Gravity studies have shown a shallow

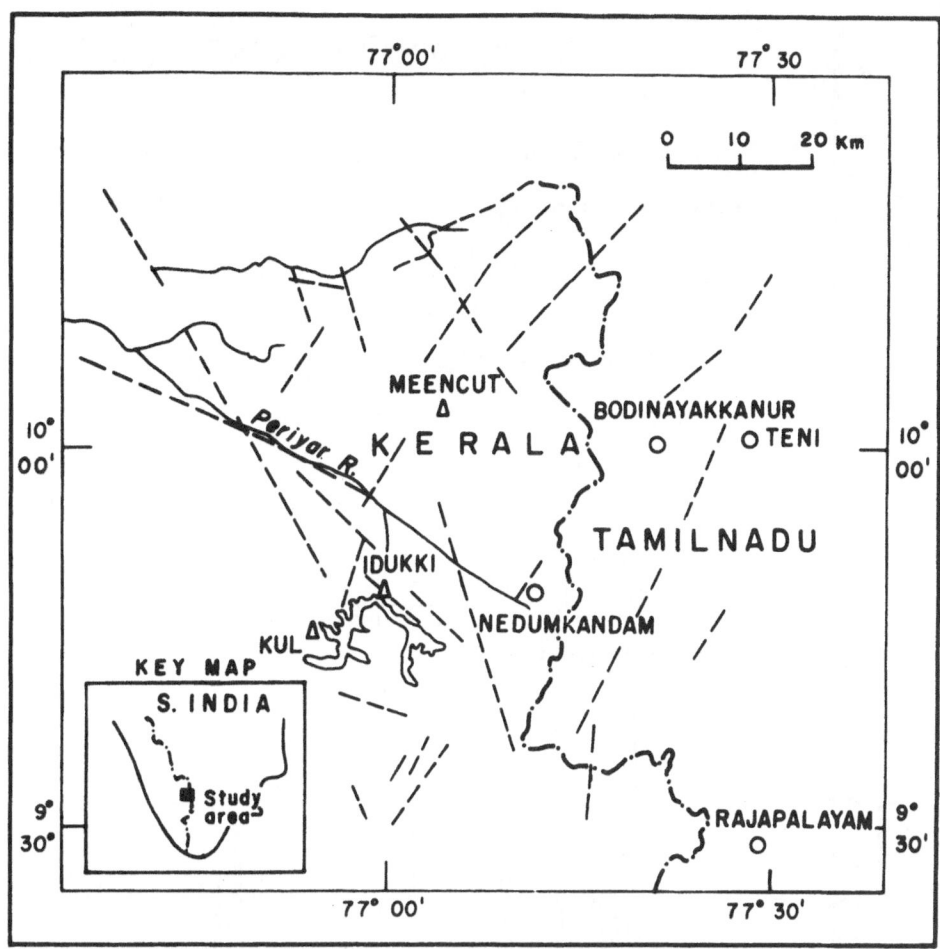

Figure 1
Important lineaments around the Nedumkandam area based on the lineament map prepared by GSI.

fault at a depth of about 4 km in this area, coinciding with the Periyar lineament (MISHRA *et al.*, 1989). This lineament passes through the epicentral area and continues up to 100 km in a northwest direction, where the Periyar River takes a E-W turn along Alwaye (West Coast). The regional trend of the fold axis in the area is NW-SE (KRISHNAN, 1953, 1966). The lineament fabric in the region shows two prominent sets, i.e., NNW-SSE and NNE-SSW. The NNW-SSE trending lineaments at many locations show dextral (right lateral) strike-slip components, indicating shear fracture domain.

Seismicity around the Idukki Reservoir (Jan. 1971–May 1988)

After the damaging earthquake in 1967 near the Koyna reservoir, situated about 1600 km north of Idukki, also near the west coast, it was realized that the filling

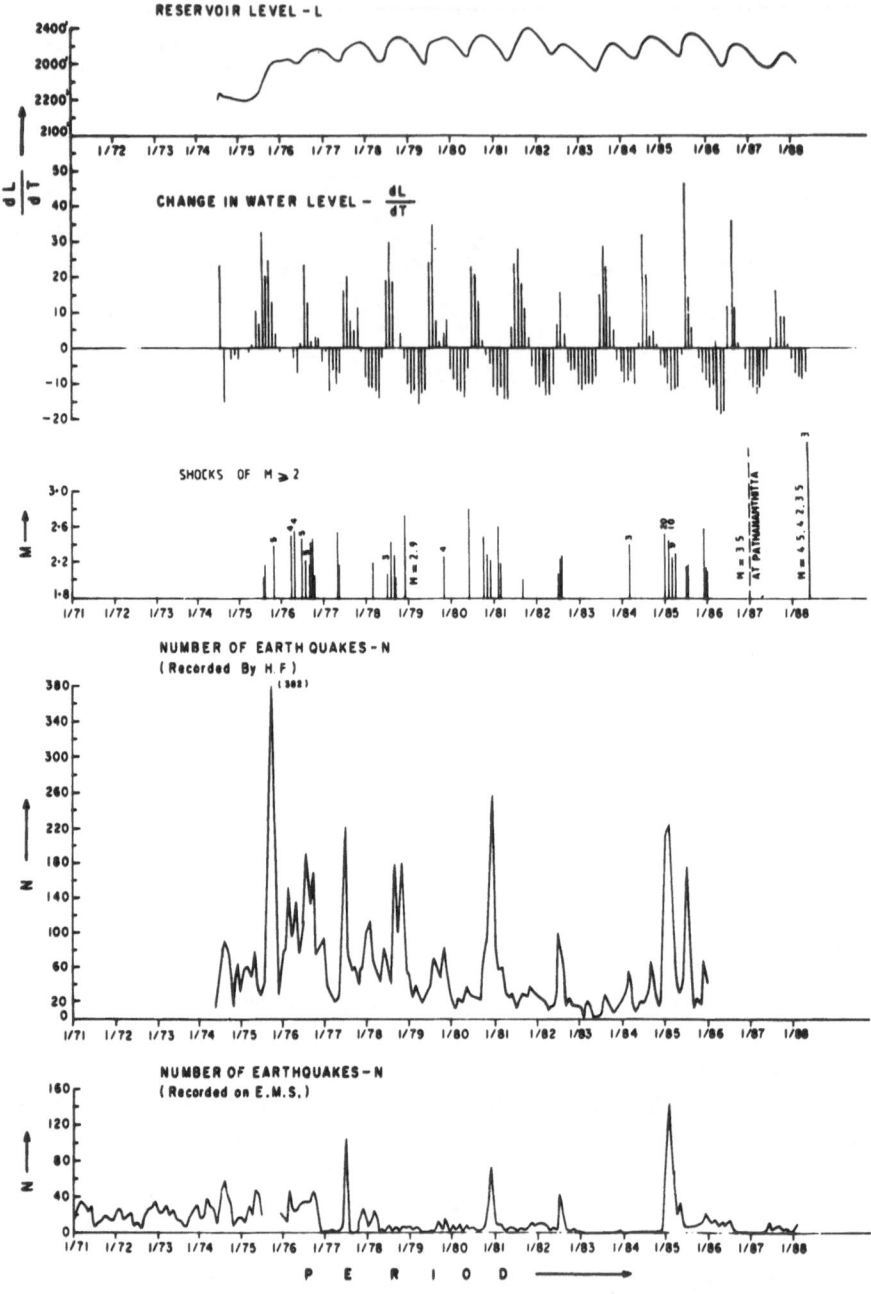

Figure 2
Idukki reservoir level, monthly change in water level, magnitude of shocks, number of shocks recorded at high frequency as well as electromagnetic seismographs (prepared by KSEB).

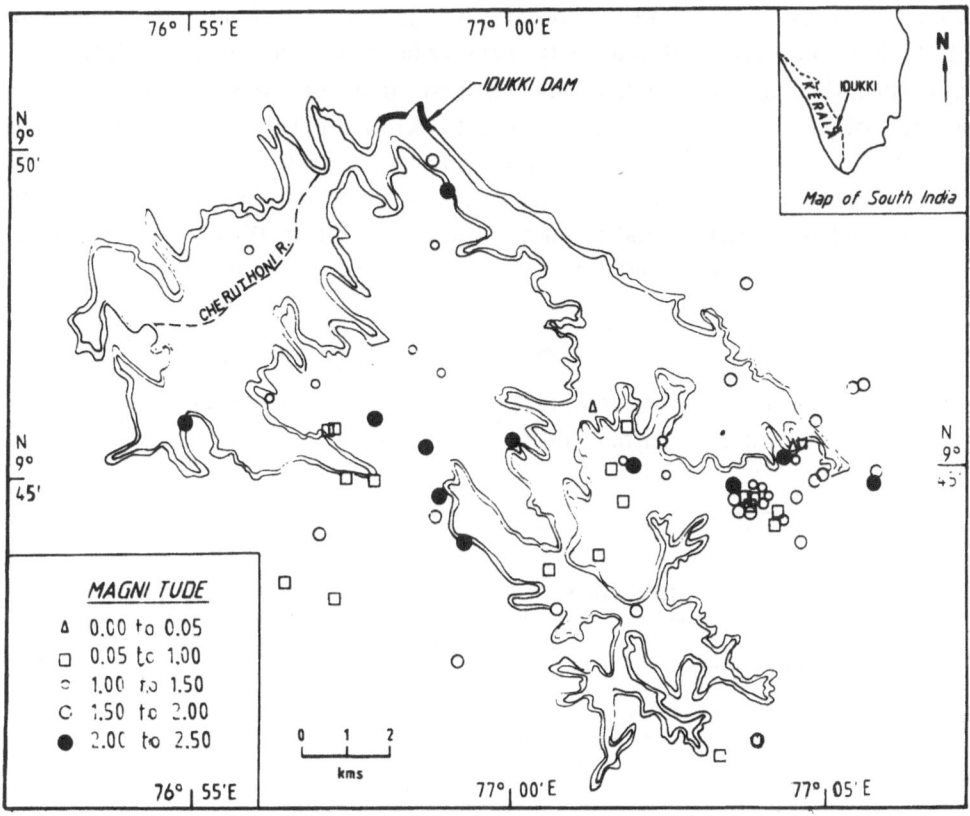

Figure 3
Some of the epicenters which could be located by the Kerala State Electricity Board during 1974–87
near the Idukki reservoir (prepared by KSEB).

of the Idukki reservoir should be done in a controlled way and seismicity should be monitored with a network of six seismic stations around the reservoir. As such the filling was accomplished slowly for a lengthy period of 7 years from 1974 to 1981, there was a provision for two large size outlets in the adjacent Cheruthoni Dam to release water when the rate of rise of the reservoir level exceeded the predetermined value. Though there was an increase in seismic activity in the form of minor tremors, no large earthquake occurred after the initial impoundment. This increase in seismicity after the impounding of the reservoir has been ascribed to an induced phenomenon (GUHA et al., 1981) from the viewpoint of nearness to the reservoir, an increase of seismicity immediately after the filling and bursts of seismicity following a high rate of loading.

The Idukki seismological network in operation since 1971 has recorded a large number of mild shocks in the project area. However, only a few of these shocks were locally felt. Figure 2 shows reservoir levels and seismicity. It is seen that prior to impoundment the electromagnetic seismograph at Idukki recorded on average about 20 shocks per month, or more than 200 per year. During impoundment (1974–81) four bursts of seismicity were noticed in October 1975, July 1977, December 1980 and Jan.–Mar., 1985. The first burst of seismicity occurred after a loading of 26 m in three months, August–October (11 m in Aug.) 1975. The high-frequency seismometer, being more sensitive, recorded even smaller shocks, hence, the larger number. Figure 3 illustrates the epicenters of some of the shocks which could be located in and around the reservoir during 1974–87. However, by no means it is a complete picture as the monitoring was not uniform.

Earthquake Sequence in Nedumkandam–Kallar Area (June–September 1988)

After the magnitude 4.5 earthquake on June 7, 1988, 20 km east of Idukki reservoir a network of 5 seismic stations was operated around the epicentral area

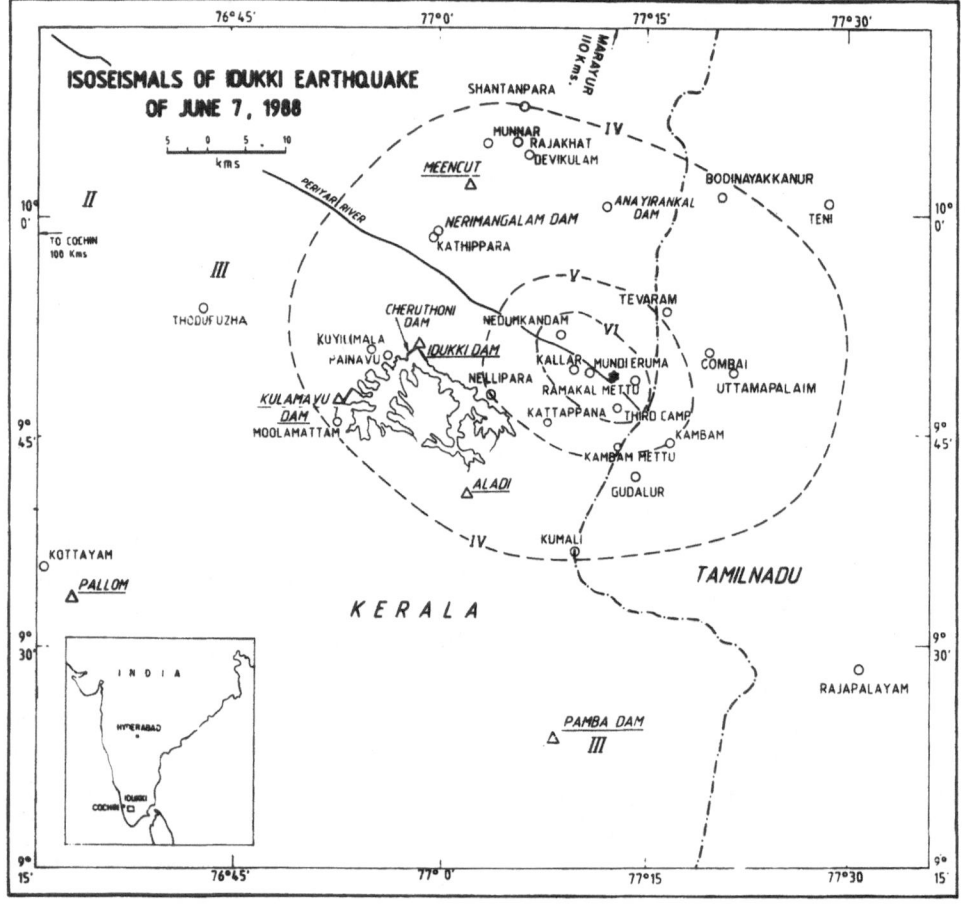

Figure 4
Isoseismal map of the June 7, 1988 earthquake.

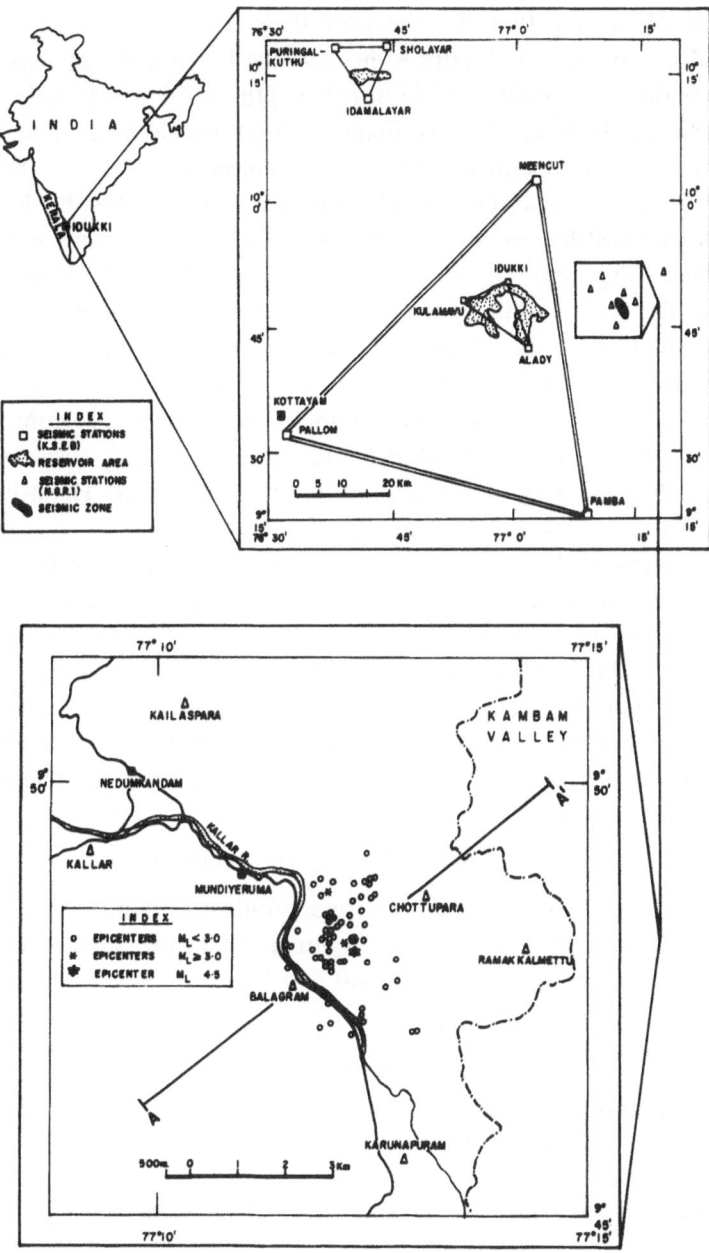

Figure 5
Idukki, Idamalayar and Nedumkandam networks of seismic stations and epicenters located in the
Nedumkandam-Kallar area from June 21 onwards.

near Nedumkandam from June 20, 1988 until the end of this sequence which lasted for three and a half months. Figure 4 displays the isoseismal map based on field studies. From the isoseismals a main shock depth of 5 km was obtained, using KARNIK's formula (1969, pp. 29–32). Figure 5 shows the seismicity and network of seismic stations which operated close to the epicentral area by the National Geophysical Research Institute (NGRI). The network run by the Kerala State Electricity Board (KSEB) around Idukki and Idamalayar Reservoirs is also shown. The instruments deployed by NGRI were Portacorders RV 320B and L-4C seismometers. The recording was done at a speed of 120 mm/min on smoked paper. Radio time signals from the National Physical Laboratory (NPL), Delhi were electronically impinged directly on records to produce a time accuracy of 10 milliseconds. The seismograms were read under a microscope, providing a reading accuracy of 0.10 sec. In the KSEB network, sensitive S-500 seismometers and Portacorders were operating at Poringalkuthu and Idamalayar. The other stations are equipped with less sensitive seismographs.

Within 24 hours of the main shock of the sequence there were two widely felt aftershocks—the first on the same night at 8.51 p.m. and the second at 8.34 a.m. on June 8, 1988. There were 26 other minor shocks felt during the first three days. Subsequently, two shocks were felt on June 21 and one each on August 26 and 27. The three widely felt earthquakes produced *MD* magnitudes of 4.5 (main shock), 4.2 and 3.5 (aftershocks), as determined from the Hyderabad Seismograph Station situated approximately 1000 km north, using duration 'D' on Benioff Seismographs. The formula used for regional shocks is,

$$MD = -0.74 + 1.67 \log D + 0.0009\Delta$$

where, Δ is the distance in kilometers. This magnitude was standardized against the M_S magnitude for teleseismic events. For regional shocks this value nearly matches the Richter scale magnitude M_L. For local shocks recorded on microearthquake recorders, we used the formula given by GUPTA *et al.* (1980),

$$MD = -2.44 + 2.61 \log D$$

where D is the duration in seconds.

Epicenters and Depth of Tremors in Nedumkandam-Kallar Area

Approximately 120 tremors were recorded from June 20 to September 28, 1988, of which two were of magnitude >3 (3.7 and 3.0), four of magnitude 2.0 to 2.9 and the rest were less than 2. Figure 6 shows the daily number of tremors recorded by the NGRI and Idukki networks, while Figure 7 shows the cumulative number of shocks recorded by the NGRI network. Sixty seven tremors down to −2.4 magnitude (duration 1 sec) were located using the HYPO-71 computer program of LEE and LAHR (1972). Most of the shocks of magnitude −0.2 (duration 7 sec) have been

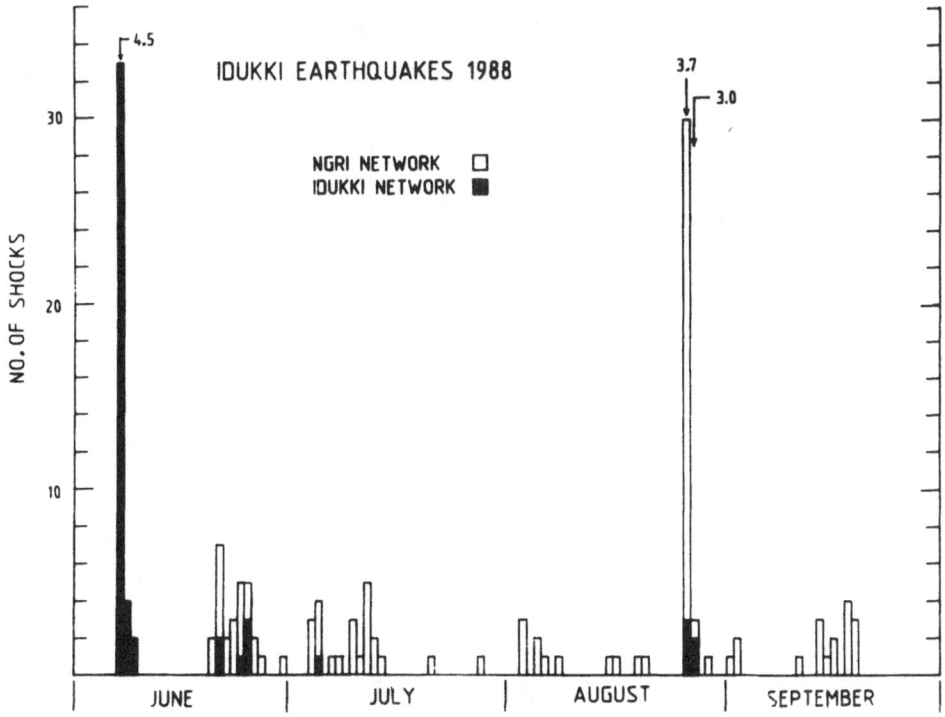

Figure 6
Daily number of shocks in the Nedumkandam-Kallar area.

located. The average root mean square (rms) of the travel-time residuals is 0.1 sec. The average horizontal error (ERH) and error in depth (ERZ) obtained for the tremors are less than 500 meters. The hypocenters are confined to a 4 km × 3 km zone along the Kallar river south-east of Nedumkandam (Fig. 5). The main event of magnitude 4.5 also occurred in this zone. The focal depths of the tremors are less than 5 km. Figure 8 shows the depth distribution along A-A' profile (Fig. 5) in SW-NE direction perpendicular to the Kallar river. All tremors are confined to a depth of 2.5 to 5 km. The depth section indicates the steep dip of the fault.

After many a trial and error, the velocity model used for these tremors, which produced the least errors, was the simplified model taken from the DSS profile over Kavali-Udipi (KAILA *et al.*, 1979), some 400 km north of the area.

Velocity Model

Velocity	Depth	V_p/V_s ratio
5.40	0.0	
5.68	2.0	
6.50	23.0	1.73
7.15	40.0	

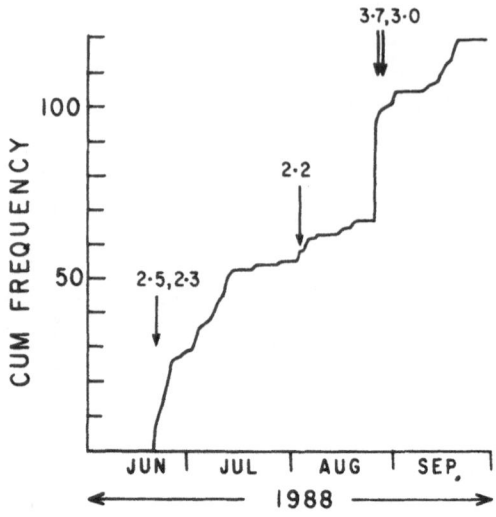

Figure 7
Cumulative frequency of shocks at the Nedumkandam-Kallar area from June 21 onwards.

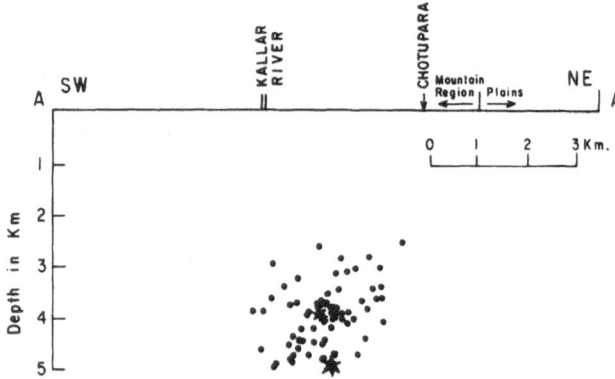

Figure 8
Focal depths of shocks in the Nedumkandam-Kallar area. The main shock is denoted by a bigger star.
Small stars indicate a shock of magnitude 3.

b Value

The b value in the frequency-magnitude relationship given by GUTENBERG and
RICHTER (1956) was determined for the earthquakes which occurred during June–
September 1988, using a least square method. A b value of 0.59 (Fig. 9) was
obtained, which is more representative of shield seismicity rather than RIS. The
magnitude ratio of the largest aftershock ($M = 4.2$) to the main shock ($M = 4.5$) is
0.93, which is high. Low b-value and high magnitude ratio are not characteristic of
RIS.

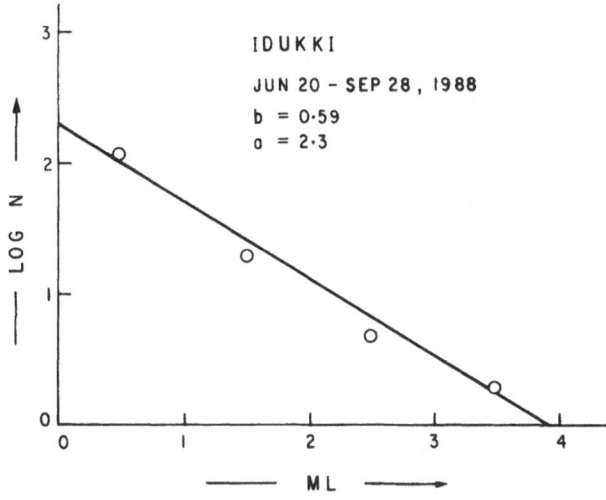

Figure 9
The *b* value from the plot of log frequency V_s magnitude.

Composite Fault Plane Solution

Composite fault plane solution (CFPS) was made for events which occurred during June 20 to September 28, 1988. Only P-wave first motions were used. The P-wave first motions were plotted on the lower hemisphere of an equal area projection (Fig. 10). There are only 12 inconsistent observations. In Figure 10 there are two vertical nodal planes in NW-SE and NE-SW directions which indicate strike-slip faulting. The movement is right-lateral along the NW-SE plane which is considered to be the fault plane from geological evidence.

Discussion

The seismic activity which started 20 km east of Idukki reservoir in Kerala was monitored with a network of five seismic stations during June 20 to September 28, 1988, after which the seismicity ceased. With a close network of seismic stations it was possible to locate fairly accurately the epicenters of these microearthquakes. The epicenters are found to lie in an area of 4 km × 3 km along the Kallar River. The depths are very shallow (less than 5 km) and are distributed in the 2.5–5 km zone. The epicenters are related to a NW-SE trending fault along the Kallar River which is a continuation of the Periyar River lineament in the area inferred from gravity and Landsat imagery. This view is also supported by the fault plane solution obtained for these microearthquakes. The CFPS suggests strike-slip faulting along a NW-SE trending nodal plane. The right-lateral movement along this fault

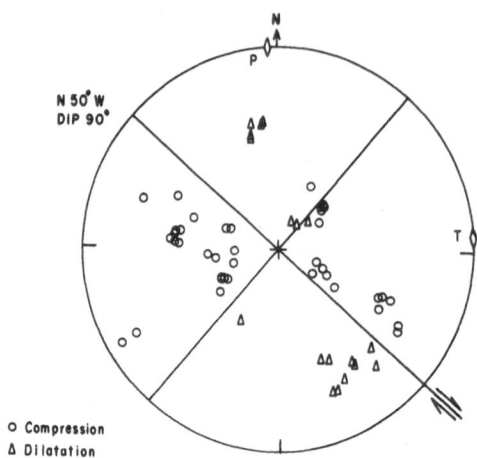

Figure 10
The composite fault-plane solution. First motion directions are plotted on the lower hemisphere of equal area projection.

matches the inferred movement along the NW-SE trending lineaments in the area from satellite imagery. Seismicity was intense for the first three days and later subsided. As the seismic network only became operational 13 days after the first felt shock of magnitude 4.5 on June 7, 1988, the epicenters of the shocks between the 7th to the 20th June could not be located. The felt reports of June 7, 1988 main earthquake indicate the epicenter to be near Mundieruma, which is southeast of Nedumkandam (situated 25 km east of the Idukki Dam or 20 km from the Idukki reservoir). The explosion-like sound reported by people around Mundieruma and Ramakalmettu also indicates nearness to the source.

Except for a few earthquakes in the historical past of southern India, the Idukki area is not seismically active. However, there were microearthquakes in the reservoir area which increased after the impoundment of the Idukki reservoir in 1974. The maximum magnitude, as reported by GUHA (1981), was around 3.0. This increased seismicity has been attributed to the reservoir impounding due to the nearness of the reservoir, the increase of seismicity immediately after the start of filling and bursts of seismicity which have occurred for several years after high rates of loading. Seismicity increased soon after the filling started in 1974. Five pronounced shocks ($M > 2$) were reported in the next year, 1975. Seismicity further increased in 1976 with the increase of storage when 22 pronounced shocks occurred. The total number of recorded shocks was more than one thousand during this initial period of impounding. There were 382 shocks recorded just in the month of October 1975. Thereafter, seismicity has continued at a slightly lower level, nonetheless, there were bursts of seismicity in 1977 and 1980 following the peak levels. The reservoir was filled to capacity during the rainy season of 1981, however seismicity was low from that time until 1984. The seismicity increased again during 1985–86.

Reservoir-induced earthquakes are nothing but triggered tectonic earthquakes. There are only subtle differences between the normal and triggered earthquakes in the form of their seismic characteristics (GUPTA and RASTOGI, 1976; GUPTA, 1992). The seismic characteristics of the Nedumkandam-Kallar sequence are normal. The reservoir-induced earthquakes are believed to be triggered by increased pore pressure, either due to the weight of the water column of the reservoir or due to water percolation, i.e., diffusion. The incremental stress due to the water load will be minor at a distance of about 20 km. Increased pore pressure is possible at these distances. An increase in pore pressure is quite often reflected in the raised water table or increased ground water flow. However, in the Nedumkandam-Kallar area we have seen no evidence of change in the ground water regime from bore-well data. Thus, from the foregoing and the following specific points we believe that the seismicity at the Nedumkandam-Kallar area from June 7 to September 28, 1988 was not induced by the Idukki reservoir.

(i) The epicentral area is about 25 km away from the Idukki Dam or approximately 20 km from the nearest reservoir boundary. Neither was a single earthquake recorded previously from this area nor was there any shock between the reservoir and the Nedumkandam-Kallar area.

(ii) The tremors began when the reservoir level at Idukki was lowest, about 10 m (28.87 feet).

(iii) There were no foreshocks prior to the main shock on June 7, 1988 and the aftershock activity in Nedumkandam-Kallar area ceased in 3 1/2 months time. Numerous foreshocks and a very large sequence of aftershocks are characteristic of reservoir-induced seismicity which is not the case for this sequence.

(iv) The b value of 0.59 for the Nedumkandam-Kallar sequence is low as compared to b values for RIS and it is similar to the normal value in Peninsular India. The magnitude ratio of the largest aftershock to the main shock is 0.93, which is high. Low b value and high magnitude ratio are characteristic of tectonic earthquakes like the Bhadrachalam earthquake sequence in Peninsular India (GUPTA et al., 1970). For reservoir-induced earthquakes both the magnitude ratio and the b value are high.

Conclusions

The seismicity sequence which followed the main shock of $M = 4.5$ in the Nedumkandam-Kallar area occurred some 20 km away from the reservoir, with no foreshocks, a faster decay of aftershocks and with no clear correlation with water levels indicating that this activity was of the swarm type associated with the release of energy from the shallow faults, as a part of the normal seismicity of the

Peninsular shield rather than induced by the Idukki reservoir. The area presents a complex tectonic setting of shallow and deep-seated faults in which crustal adjustment in the upper crust is not a new phenomenon.

Acknowledgements

The authors wish to thank Dr. Harsh K. Gupta, Director, National Geophysical Research Institute, Hyderabad for his constant encouragement and valuable advise for these studies and the permission to publish this work. Appreciation is extended to the Kerala State Electricity Board for providing the data of KSEB network of seismic stations and assistance in our field work. Ms Narendra Kumar, C. Satyamurthy, I. P. Raju, A. Nageswara Rao and B. Ramamurthy operated the seismic stations and analyzed the data. Ms. Jaffar Ali and M. Kranti Kumar traced the drawings.

REFERENCES

GUHA, S. K., GOSAVI, P. D., VARMA, M. M., AGARWAL, S. P., PADALE, J. G. and MARVADI, S. C. (1970), *Recent Seismic Disturbances in the Shivajisagar Lake Area of the Koyna H. E. Project, Maharashtra, India*, Central Water and Power Research Station, Khadakwasla, Pune, pp. 56–89.

GUHA, S. K., PADALE, J. G., and GOSAVI, P. D., *Probable risk estimation due to reservoir-induced seismicity. In Dams and Earthquakes* (Institution of Civil Engineers, London 1981) pp. 297–305.

GUPTA, H. K., and RASTOGI, B. K., *Dams and Earthquakes* (Elsevier, Netherlands 1976) 227 pp.

GUPTA, H. K., *Reservoir-induced Earthquakes* (Elsevier, Netherlands 1992) 364 pp.

GUPTA, H. K., MOHAN, I., and NARAIN, H. (1970), *The Godavari Valley Earthquake Sequence of April 1969*, Bull. Seismol. Soc. Am. *60*, 601–615.

GUPTA, H. K., RAO., C. V. R. K., RASTOGI, B. K., and BHATIA, S. C. (1980), *An Investigation of Earthquakes in Koyna Region, Maharashtra for the Period October 1973 through December 1976*, Bull. Seismol. Soc. Am. *70*, 1833–1847.

GUTENBERG, B., and RICHTER, C. F. (1942), *Earthquake magnitude intensity, energy and acceleration*, B.S.S.A., *32*, 163–191; second paper, *ibid*, *46* (1956), 105–145.

KAILA, K. L., ROY CHOWDHURY, K., REDDY, P. R., KRISHNA, V. G., HARINARAYANAN, A. V., SUBBOTIN, S. I., SOLLOGUB, V. B., CHEKUNOV, A. V., KHARETCHO, G. E., LAZARENKO, M. A., and ILCHENKO, T. V., (1979). *Crustal Structure along Kavali-Udipi Profile in the Indian Peninsular Shield from Deep Seismic Sounding*, J. Geol. Soc. India *20*, 307–333.

KARNIK, V., *Seismicity of the European Area, Part 1* (Reidel, Dordrecht-Holland 1969) 364 pp.

KRISHNAN, M. S. (1953), *Structural and Tectonic History of India*, Mem. Geol. Surv. India *81*, 1–109.

KRISHNAN, M. S. (1966) *Tectonics of India*, Bull. Ind. Geophys. Union *3*, 2–36.

LEE, W. H. K., and LAHR, J. C. (1972), *A Computer Programme for Determining the Hypocenter, Magnitude and First Motion Pattern of Local Earthquakes*, U.S. Geol. Surv. Open file Rep., 100 pp.

MISHRA, D. C., SINGH, A. P., and RAO, M. B. S. V. (1989), *Idukki Earthquake and its Tectonic Implications*, Geol. Soc. India *34*, 147–151.

PADALE, J. G., and DAS, P. B. (1988), *A Note on the Recent Seismic Activity in the Idukki District, Kerala*, Central Water and Power Research Station, Pune, Report, 12 pp.

(Received September 12, 1994, revised/accepted February 10, 1995)

PAGEOPH, Vol. 145, No. 1 (1995)

0033–4553/95/010123–15$1.50 + 0.20/0

Application of Stress-pore Pressure Coupling Theory for Porous Media to the Xinfengjiang Reservoir Earthquakes

Li-Ying Shen[1] and Bao-Qi Chang[1]

Abstract —Theory of the coupling of stress-pore pressure in the saturated, elastic porous media is used in the study of the formation mechanism of the Xinfengjiang reservoir-induced earthquakes. Based on the results, it is believed that compared with the mechanism of additional stress in the vicinity of the reservoir, the mechanism of the coupling of additional stress and pore pressure may be more well-founded for the occurrence of reservoir-induced earthquakes.

Key words: Porous media, coupling theory, Xinfengjiang reservoir, induced earthquakes.

Introduction

The Xinfengjian reservoir is one of the four great reservoirs in the world which has induced an earthquake $M > 6.0$. The purpose of this paper is to discuss the effects of the reservoir water on the induced earthquake, using the realistic conditions of the Xinfengjiang reservoir. Calculation for the stress seepage field of water pressure in the Xinfengjiang reservoir has been made respectively, but it would be more significant to consider the induced effect of water under the coupling of stress and pore pressure in the more complicated and saturated, elastic porous media. This concept has been used by Bell and Nur (1978) to study the Oroville Lake reservoir. Roeloffs (1986) considered three types of faults and used the annual variation of water level for the study of the Oroville and Lake Mead reservoirs, etc. This concept is also adopted in this paper to study the Xinfengjiang reservoir earthquakes. Except for the effect of the type of fault on the coupling, the effect of the high dense base in the Xinfengjiang reservoir area and the actual variation of reservoir water levels with time are considered. This concept is more convincing for the explanation of the occurrence of the Xinfengjiang reservoir-induced earthquakes.

---

[1] Seismological Bureau of Guangdong Province, 81 Xianlie R.C., Guangzhou, P.R. China.

Theory and Equations for Calculation

RICE and CLEARY (1970) further developed Biot's linear pseudo-static elastic theory related to fluid-saturated porous media. They considered the porous medium as a compressible constitutent, and used new physical parameters, V_u (undrained Poisson's ratio) and B (Skempton coefficient), instead of the physical constants, H and R in Biot's theory. Based on these parameters, constitutional equations and field equations for the stress-pore pressure coupling on porous media were derived as follows:

Constitutional equations:

$$2G\varepsilon_{ij} = \sigma_{ij} - \frac{v}{1+v}\sigma_{kk}\delta_{ij} + \frac{3(v_u - v)}{B(1+v)(1+v_u)}p\delta_{ij} \tag{1}$$

$$m - m_0 = \frac{3\rho_0(v_u - v)}{2GB(1+v)(1+v_u)}\left[\sigma_{kk} + \frac{3}{B}\rho\right] \tag{2}$$

where,

ε_{ij} = strain component of the porous medium,

σ_{ij} = stress component of the porous medium,

p = pore pressure of the fluid,

$m - m_0$ = difference of mass constituents of the fluid in a unit volume of the porous medium,

ρ_0 = density of the fluid in the pores,

δ_{ij} = Kronecker operator,

$G\ V$ = shear modulus and Poisson's ratio of the medium under drained conditions, respectively.

Field equations for the deformation-pore pressure coupling:

$$(\lambda + G)\frac{\partial \varepsilon_{kk}}{\partial x_i} + G\nabla^2 u_i - \frac{3(v_u - v)}{B(1+v_u)(1-2v)}\frac{\partial p}{\partial x_i} = 0 \tag{3}$$

$$K\nabla^2 p = \frac{3(v_u - v)}{2GB(1+v)(1+v_u)}\frac{\partial}{\partial t}\left(\sigma_{kk} + \frac{3}{B}p\right) \tag{4}$$

where,

λ = Lamé constant,

$K = k/\mu$, in which k is the seepage rate, μ is the viscosity coefficient of the fluid.

In order to utilize the displacement method conveniently in the finite element method, eq. (4) will be expressed in terms of displacement and pore pressure.

$$K\nabla^2 p = \frac{3(v_u - v)}{2GB(1+v)(1+v_u)}\frac{\partial}{\partial t}\left[\frac{E}{1-v}\varepsilon_{kk} + \frac{3(1+v)(1-2v)}{B(1+v_u)(1-2v)}p\right] \tag{5}$$

where,

$\varepsilon_{kk} = \partial u/\partial x + \partial u/\partial y + \partial w/\partial z$,

E = elastic modulus.

Using the program for the two-dimensional finite elements calculation devised by the authors, the displacements, u, v and pore pressure p are first obtained, then we derive the strain and the stress. In order to obtain the variation of strength of the fault induced by the stress and pore pressure produced by the reservoir water load, the stength increment acting on the fault plane can be given by the following equation, based on the Mohr-Coulomb criterion and the effective stress law

$$S = -\mu_1(\sigma_N + p) \pm \tau_N \tag{6}$$

where,

μ = friction coefficient,

σ_N = normal stress on the fault plane produced by reservoir water; + indicates tensile stress, − indicates compressive stress,

τ_N = shear stress on the fault plane produced by the reservoir water,

p = pore pressure; + sign indicates increase of the pressure, − sign indicates decrease of the pressure,

S = value showing the variation of strength of the fault; $S > 0$ shows the fault will be strengthened, $S < 0$ shows the fault will be weakened.

Because the shear stress directions induced by the reservoir water load are not identical in the reservoir base, the sign before τ_N in eq. (6), whether it is + or − not only depends on the features of the fault, but also depends on the magnitude of the fault dip, and cannot be considered as the same in all cases. When the fault is a strike-slip fault, τ equals zero as the model is two-dimensional.

Calculation Results

1. Effect of Water Load in Elastic Model

In order to provide a comparison with the effect of coupling, the elastic deformation produced by the water load will only be considered in the model of this paper. Assume the media in the Xinfengjiang reservoir area is elastic and homogeneous. The water load after filling in a symmetric step-wise shape is shown in Figure 1 and the maximum depth of water is 90 m. This corresponds to the pure elastic problem of the force acting on the boundary of a semi-infinite plane, after which the variation of strength of the fault will be shown by eq. (7) instead of eq. (6),

$$S = -\mu_1\sigma_N \pm \tau_N. \tag{7}$$

It can be found that, when the effect of additional stress induced by the water load is only considered, although weakening will occur in part of the area, strengthening still plays the main role.

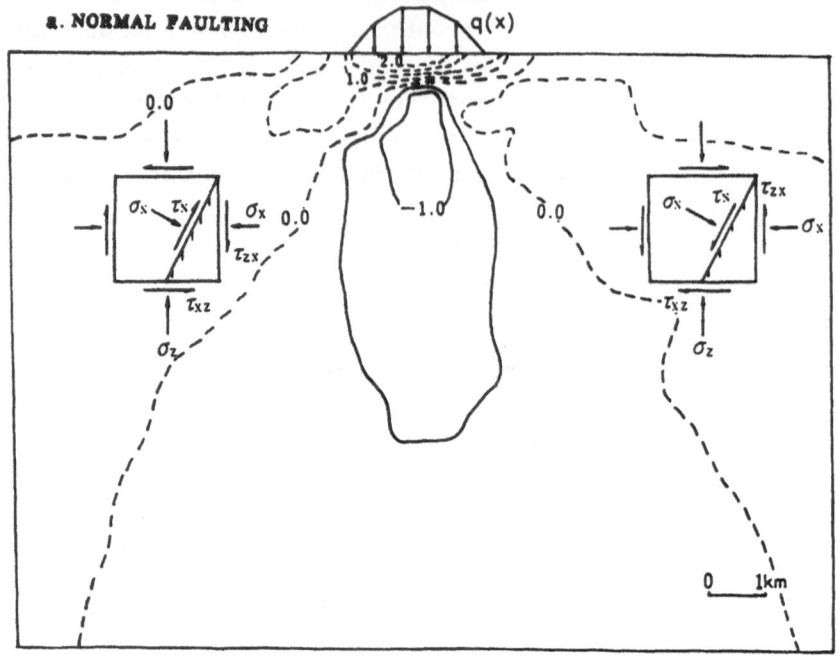

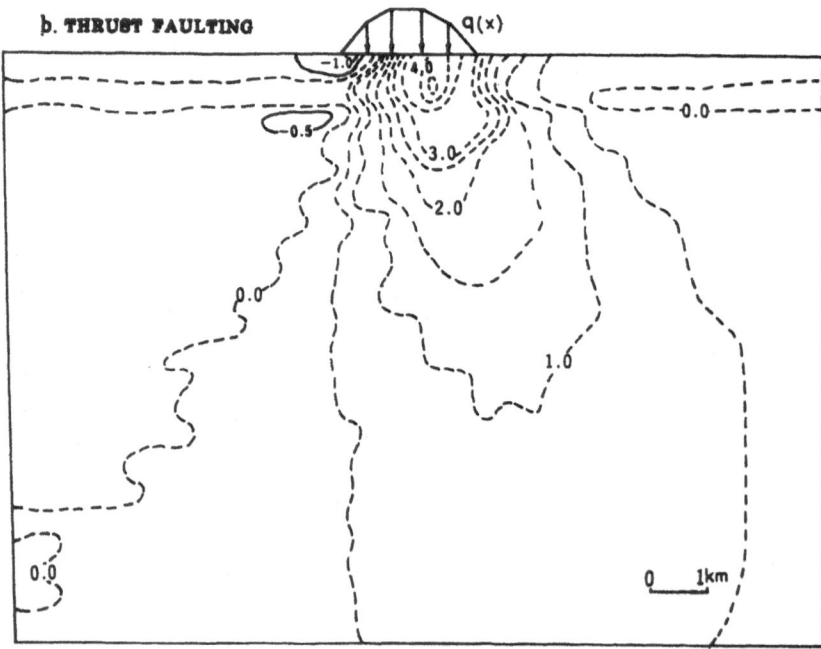

Figure 1
Variations of strength of the fault with a dip of 60° produced by the additional stress of the reservoir
water. Unit of stress is 10^{-1} MPa.

2. Application of the Homogeneous Coupling Model

We first consider the case in which the parameters of rock mechanics and those of fluid seepage in the entire reservoir base are all constant (from Roeloff's data), i.e., $v = 0.25$, $v_u = 0.30$, $B = 0.8$, $k = 10^{-6} \, \text{m}^2$, μ is 10^{-2} poise, μ_I and the water load are the same as mentioned above. Because the model is homogeneous, the pore pressure extends outwards and downward in the reservoir area uniformly in the shape of a semi-circle (Fig. 2).

Figure 3a shows the variation of strength induced by the homogeneous coupling effect of a fault with a dip of 60° (the angle with x axis as one side in the negative direction). In the case of a normal fault (Fig. 3a), reservoir filling caused weakening to occur below the reservoir area with a maximum value of 0.47 MPa, which attenuates outward in the shape of an ellipse. Weakening attenuates more quickly in the upper and more slowly in the lower. At a depth of 4 km, the weakened value can reach 0.2 MPa value. In the case of a strike-slip fault (Fig. 3b) the entire reservoir area will be weakened. The maximum weakened value is above 0.29 MPa. Figure 3c demonstrates that in case of a reverse fault, weakening will occur throughout most of the reservoir base due to coupling, and strengthening will occur in a small part only, because the strengthened values are small. In the case of a strike-slip fault with a dip of 90°, the coupling effect induced by filling will weaken

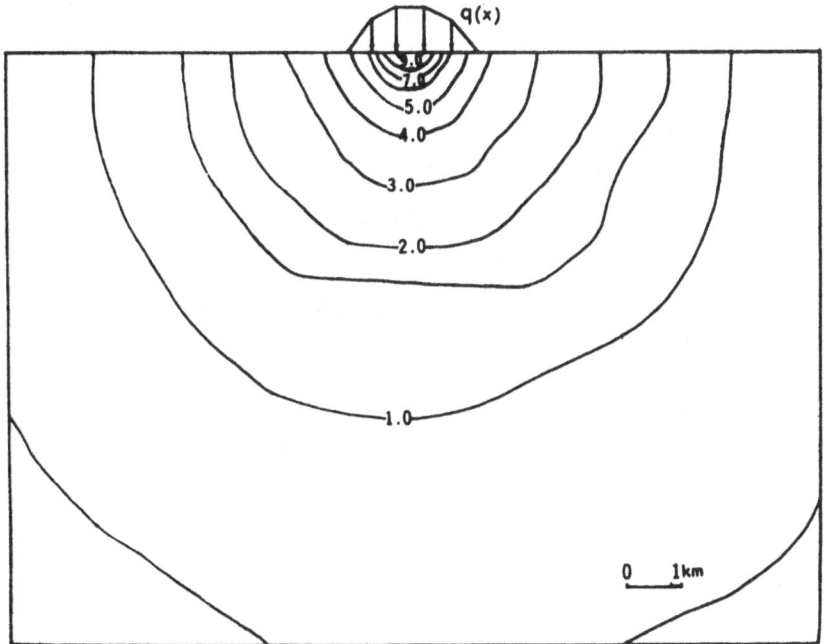

Figure 2
Distribution of pore pressure in the homogeneous coupling model.

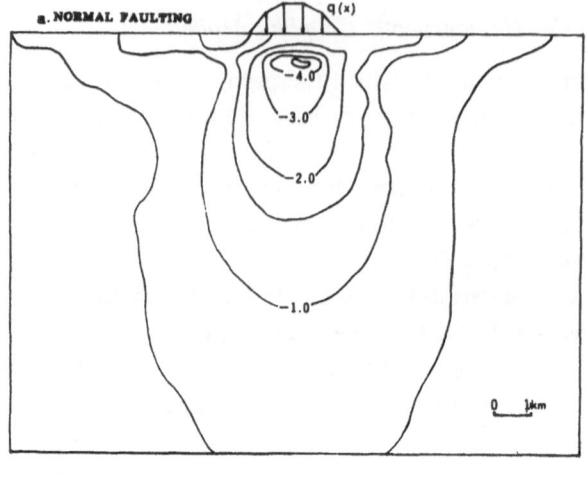

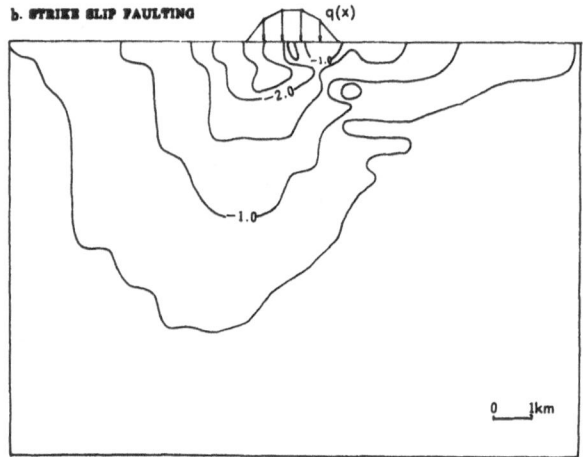

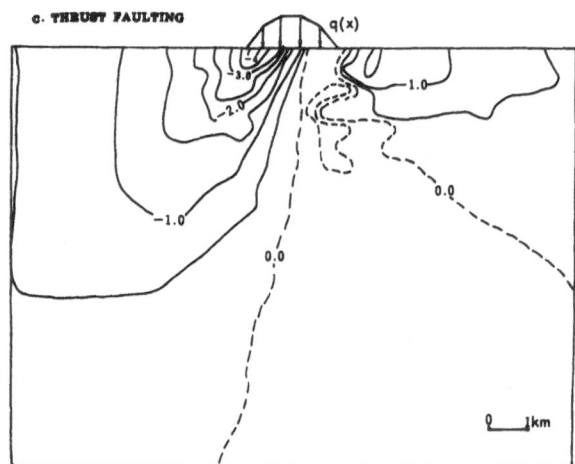

Figure 3
Variation of strength on the plane of a fault with a dip of 60° in the homogeneous coupling model.

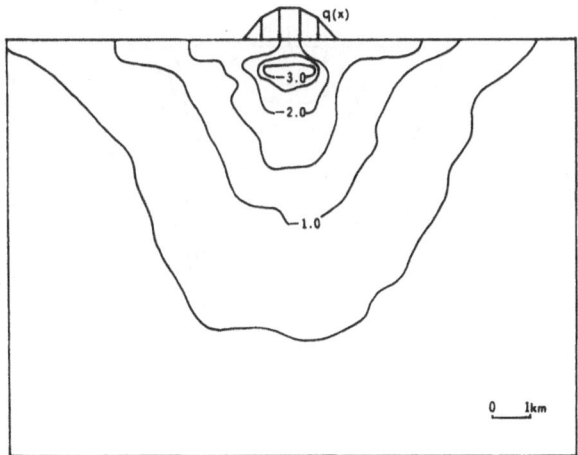

Figure 4
Variation of strength on the vertical strike-slip plane of the fault (homogeneous coupling model).

the entire reservoir base (Fig. 4). Calculation results of the homogeneous coupling model express that the effect of coupling of stress-pore pressure can weaken the strength of the fault in the reservoir region. These results also indicate that normal faults and strike-slip faults are likely to be weaker than reverse faults and that faults with a steep dip angle are also likley to be weaker than faults with a smooth dip angle.

3. Application of the Nonhomogeneous Coupling Model in the Xinfengjiang Reservoir Area

Basically, media in the Xinfengjiang reservoir area are not homogeneous. In this area, as regards the lithological characteristics, there are many muddy shales and gravy shales of the Jurassic, Cretaceous and Tertiary periods in the vicinity of the reservoir area excepting granite in the epicentral area, and in addition there exist numerous discontinuous texture planes with joints, cracks and faults in the reservoir area. Based on the data of seismic exploration, it is inferred that under the granite mass there is a high-dense rock base. As a nonhomogeneous model, parameters of the high-dense base, sandy shale in the vicinity and granite in the reservoir area, will not be identical, whereas the seepage rate of the sandy shale and high-dense base will decrease, i.e., the seepage rate of sandy shale is lower than that of granite by one order of magnitude ($k = 10^{-7}\text{m}^2$), and in turn that of the high-dense base is lower than that of sandy shale by one order of magnitude ($k = 10^{-8}\text{m}^2$). B value of the high-dense base is taken as 0.9, and that of the sandy shale is 0.8 for granite rock mass, $v = 0.22$, $v_u = 0.24$ and those for the high-dense base and sandy shale, $v = 0.25$, $v_u = 0.30$. Figure 5 shows the distribution of pore pressure in the nonhomogeneous coupling model. Compared with Figure 2, the distribution is irregular

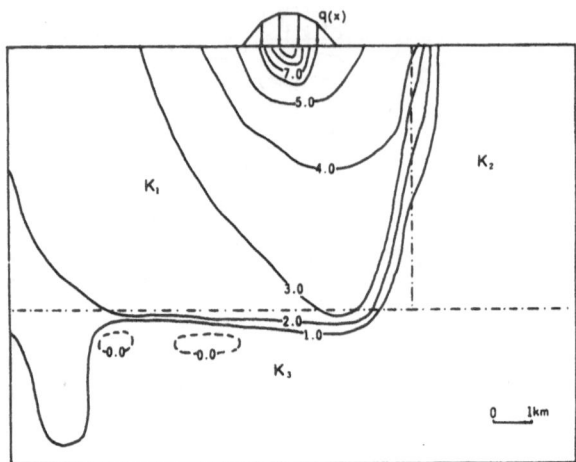

Figure 5
Distribution of pore pressure in the nonhomogeneous coupling model.

due to the existence of the high-dense base and sandy shale in the vicinity considered, and the pore pressure has a tendency to turn to the rock mass with a low seepage not far from the water domain. Pore pressure attenuates more quickly at the interface between granite and the other two kinds of rock. In the granite mass, the pore pressure value is relatively higher than that in the homogeneous model.

Figure 6 illustrates the variation of strength in a fault with a dip of 60° in the nonhomogeneous model. When compared with the homogeneous model (Fig. 3), the strengthened area is developed, while the weakened area is relatively decreased. However, its not difficult to find that in the weakened area where granite exists, although the maximum weakened value has not increased, the range of weakening of the same order increases. It can be determined that when rock of low permeability and the highly dense base exist in the vicinity of the reservoir area, the strength of the reservoir base will be weakened. When a surface fault does not reach the reservoir base, the weakened strength of the base will be higher than when a fault penetrates the base.

4. Application of the Nonhomogeneous Coupling Model under the Effect of the Filling Time History in the Xinfengjiang Reservoir

In the Xinfengjiang area, not only is the distribution of lithological characters nonhomogeneous, but also many groups of discontinuous texture plane exist (Fig. 7), such as the fault of the NE-NNE strike which is relatively large and extends for a long distance. Two group of tectonics of the NNW and NNE strike are rather small, but they are distributed closely in the epicentral area and are comparatively

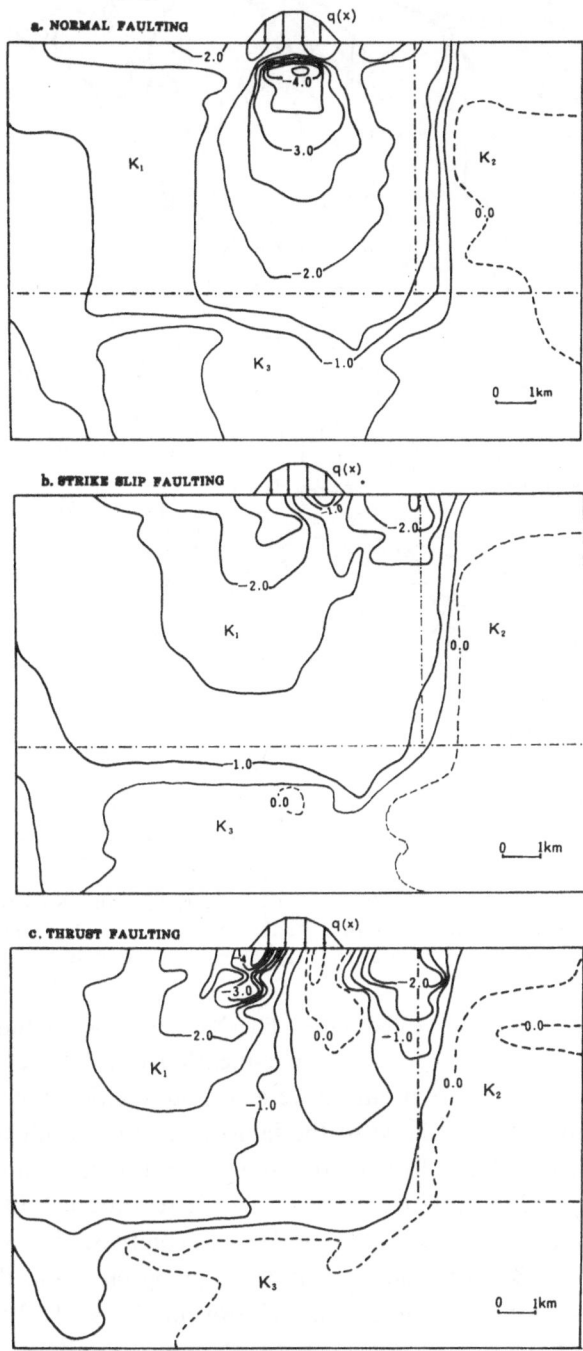

Figure 6
Variation of strength on the plane of a fault with a dip of 60° in the nonhomogeneous coupling model.

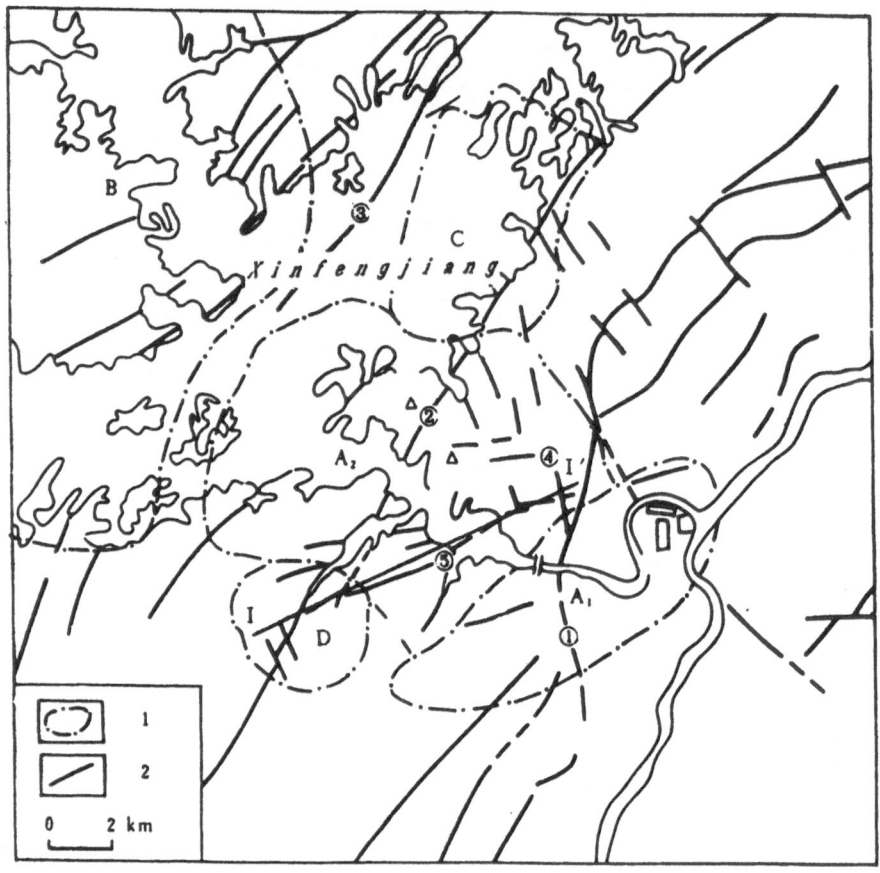

Figure 7
Geological, tectonic structures and distribution of epicenters in the Xinfengjiang reservoir area.
1. Earthquake partition; 2. Fault 1) Heyuan falult, 2) Renzishi fault, 3) Dengta-Kejiashui fault,
4) Shuangtang fault and 5) Takeng-Huluao compression zone.

active, thus they are causative tectonics. The Xinfengjiang earthquakes are closely related to the distribution of the above-mentioned NNW and NNE tectonic structures in space distribution, and have a definite relationship with the filling of the reservoir in time (also related to the faults). In order to illustrate the relationship between the coupling effect of stress-pore pressure and water level as well as earthquakes, a profile is selected in accord with realistic condition (Fig. 8). The profile is situated along the NEE Takeng-Huluao compressed zone (I-I in Fig. 7) in which the NNW faults exist, such as the Shuangtang fault and the NNE Renzishi fault. Parameters used for model are shown in Table 1. In the calculation, distribution of the reservoir water has been mentioned as above and the maximum depth of the water is 90 m. Peak and trough values of the water level (monthly average) from October 20, 1959 (initial filling) to May 1977 are input according to the realistic time intervals, after which the pore pressure, stress and variation of

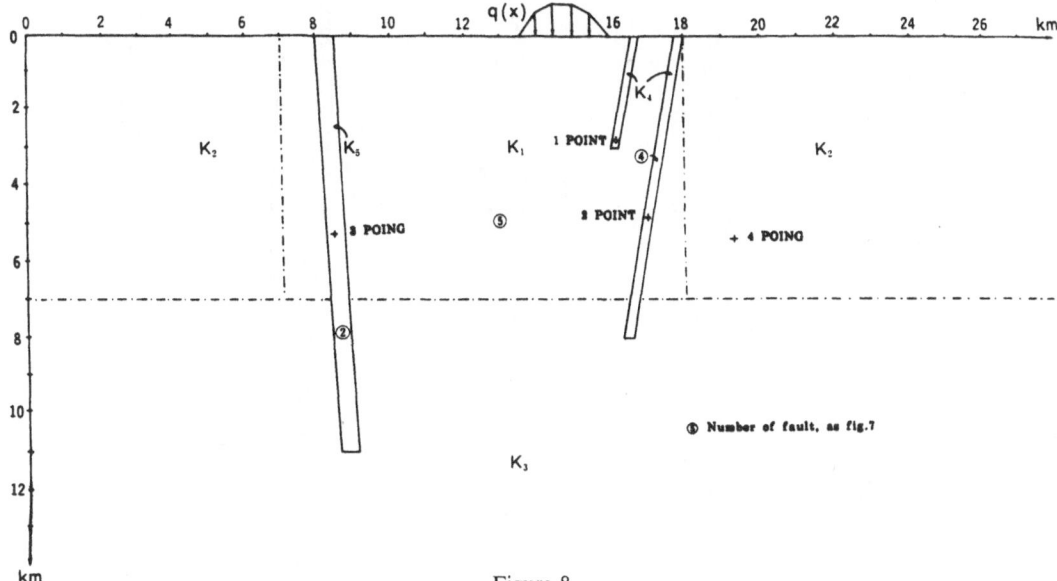

Figure 8
Profile I–I′ and locations related for calculation.

fault strength in each time segment are calculated. The results are presented in Figure 9 in which the location of point 2 is located in the vicinity of the source of the main shock of $M = 6.1$. The calculation results are summarized as follows:

i) In the initial period of filling, the weakened values in curves 1, 2 and 3 increase with the rapid increase of the reservoir water level. After the first high level, the water level fluctuates at approximately 100 m and the weakened value fluctuates in the neighborhood of a certain value, depending on its location.

Table 1

Parameter of the coupling model for the profile I–I′ of the Xinfengjian reservoir area

Name of fault	G MPa	v	v_u	B	μ_1	α	K/μ cm/MPa/day	Fault type
Shuangtang fault etc.	39216	0.22	0.24	0.8	0.5	−70	8640	Strike slip
Renzishi fault	39216	0.22	0.24	0.8	0.6	80	3000	Reverse
Reservoir base	49020	0.22	0.24	0.8	0	0	5000	—
Highly dense base	68627	0.25	0.30	0.9	0	0	0.64	—
Sandy shale	49020	0.25	0.30	0.8	0	0	816.4	—

*α is an angle between the fault and X axis.

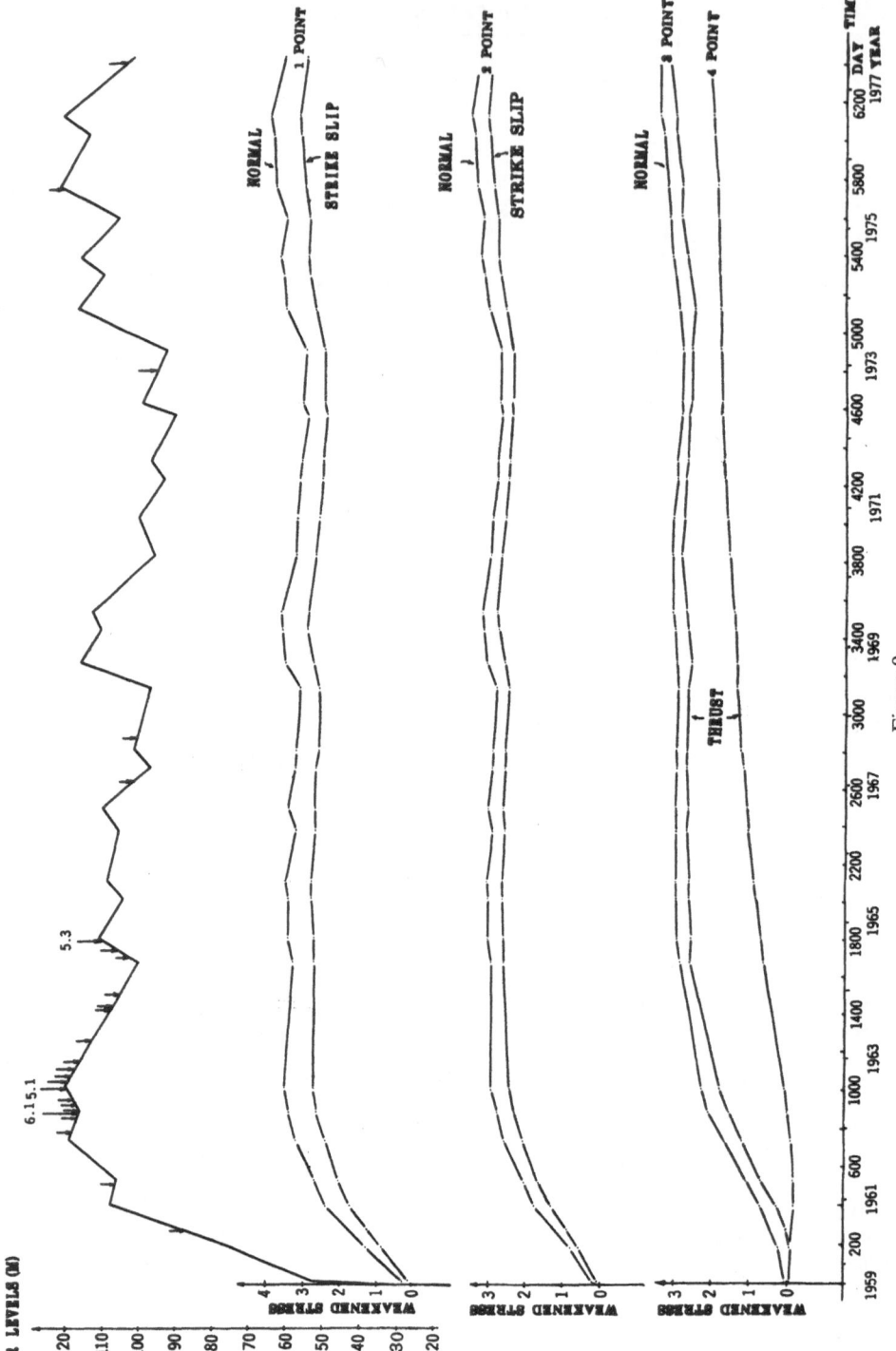

Figure 9

Relationship between the variation of water level, weakened values and earthquakes ($M_s > 4.0$).

ii) When the water level increases or decreases quickly and then turns in the reverse direction slightly, the weakened value of certain points does not change at once with the water level, but increases or decreases continuously. The amplitude of increase or decrease is rather small.

iii) As the distance from the water domain is different and the effect of the fault is also different, i.e., points near the water domain will respond quickly and points distant from the water domain will respond slowly. When the fault features are different, points located at the normal fault will respond more quickly than those at the reverse fault and sometimes the weakened value of the point located at the reverse fault is correlated negatively with the water level.

This results evidences that when the first peak water level occurred, the variation of strength in a certain range under the reservoir reached an "extreme value," subsequently, the range of variation will not be exceedingly large with the change of water level. At the same time as the velocity and amount of filling and release of water were different, the location and feature of faults were different also, therefore the relationship between the variation of strength and that of the water level became more complicated.

Discussion

1. Relationship of the Variation of Fault Strength in the Xinfengjiang Reservoir Area and the Occurrence of Earthquakes

In the regional tectonic stress field, although the two NNW and NEE tectonic systems in the area were in a state likely to loose their stability (compared with the other systems in different directions), if the reservoir would not have been built, earthquakes would not occur temporarily (or permanently) because these two tectonic systems with a steep dip angle are located in the valley region and near the edge of the reservoir. On the other hand, these two systems are under the action of strike slip and normal fault, therefore they could be easily weakened after filling. From the calculation results, the main fault in the area (the Heyuan fault) would be weakened after filling. However, owing to its low dip angle and unconnected state with the reservoir directly, its weakening capacity is comparatively small.

2. Relationship of the High-dense Base and the Xinfengjiang Reservoir Earthquakes

The Xinfengjiang earthquakes, which mainly occurred in the rock mass of granite, were also related to the high-dense base buried under the rock mass of granite in the reservoir area. Calculation of the coupling of stress-pore pressure proves the above condition: when the highly dense base exists, the weakening range of above 0.1 MPa will be more than that of the uniform reservoir base for 1–2 times (Fig. 6), that is to say, the weakening range of higher order is enlarged. The larger the range of higher weakened value, the more open is the occurrence of induced earthquakes, because the release of the main shock of $M = 6.1$ needs a certain volume of accumulated energy. From the above-mentioned, it can be inferred that the existence of a highly dense base not only accelerates the Xinfengjiang reservoir-induced earthquake, but also increases the intensity of earthquakes in that area. In addition, it should be pointed out that the existence of sandy shale around the granite mass is one of the factors necessary to trigger the occurrence of induced earthquakes.

3. Relationship of Filling-time History of the Xinfengjiang and the Induced Earthquakes

In the initial filling period (before 1965) the correlation between the Xinfengjiang earthquakes and the water level was comparatively good. The main shock of $M = 6.1$, the strong aftershock of $M = 5.1$ and numerous earthquakes of $M > 4$ occurred during the period of the first high water level (> 110 m), of which the main shock occurred 150 days after the occurrence of the first peak level value (113 m). The strong aftershock of $M = 5.3$ occurred during the period in which the second highest water level arose (105.5 m). After 1965 the relationship between earthquakes and the water level was less obvious. The weakened value produced by the coupling of stress-pore pressure varies with the response velocity of the water level, due to the location and faulting conditions.

Conclusions

The Xinfengjiang reservoir earthquake was induced by many factors, but compared with the additional stress produced by the reservoir water, the coupling effect of additional stress and pore pressure is the main factor for the reservoir-induced earthquake. The coupling effect in the Xinfengjiang reservoir area not only depends on the fault feature and highly dense base in the area, but also on the loading and unloading pattern of the reservoir water. The preliminary calculation results explain why the occurrence of earthquakes in the area were related to the water level in the initial period of filling but not in the subsequent period. The cause of the main shock lagging behind the first peak water level is also explained.

REFERENCES

BELL, M. L., and NUR, A. (1978), Strength Changes due to Reservoir-induced Pore Pressure and Stress and Application to Lake Oroville, J. Geophys. Res. 83, 4469–4483.
BIOT, M. A. (1941), General Theory of Three-dimensional Consolidation, J. Appl. Phys. 12.

RICE, J. R., and CLEARY, M. P. (1970), *Some Basic Stress Diffusion Solutions for Fluid-saturated Elastic Porous Media with Compressible Constituents*, Rev. Geophys. Space Phys. *14*, 227–241.

ROELOFFS, E. A., CHO, T. F., and HAIMSON, B.C. (1986), *Stress and Pore Pressure Changes due to Annual Water Level Cycles in Seismic Reservoir*, U.S. Geol. Surv. Contract 14–08–0001–22022.

ROELOFFS, E. A. (1986), *Fault Stability Changes Induced Beneath a Reservoir with Cyclic Variations in Water Level*, U.S. Geol. Surv. Contract 14–08–0001–22022.

(Revised March 30, 1994, revised/accepted November 4, 1994)

PAGEOPH, Vol. 145, No. 1 (1995)

0033–4553/95/010139–09$1.50 + 0.20/0

Unstable Steam-water Convection as Possible Trigger to Earthquakes

V. S. PANFILOV,[1] and G. A. SOBOLEV[2]

Abstract—Convection occurs when two water reservoirs, the overlying and the underlying, are connected by a narrow channel and the fluid in the lower reservoir is heated to the stage of phase transition into steam. The laboratory study of the properties of unstable steam-water convection showed that under favourable P-T conditions the convection can be the triggering mechanism of seismicity. This type of convection causes a sudden fall of pressure in the lower reservoir and in the connecting channel, the impulsive mechanical disturbances, and cyclicity. The point of initiation of this phenomenon can be located at a depth of 5–7 km from the earth's surface with subsequent propagation of the process of instability to larger and smaller depths. This model of the natural terrestrial conditions can account for the earthquake cyclicity in the same focal zones, the rise of temperature and of the level of ground waters during earthquakes, the enhancement of seismicity while filling the water storage basins, the effect of "floating up" of hypocenters of aftershocks and the greater intervals between them.

Key words: Convection, triggered earthquakes.

Introduction

Since the experiments in Denver (PAKISER *et al.*, 1969) have shown that the pumping of fluid into bore-holes can produce seismicity, the interest in this phenomenon has steadily increased. A review of the seismicity associated with the filling of the water reservoirs has been given by GUPTA (1992).

Two mechanisms have been proposed to date, where the intrapore fluid in rocks affects the mechanical instability. The first mechanism is associated with the increase of pore pressure and the corresponding fall of effective pressure (HUBERT and RUBEY, 1959; SIMPSON, 1986). The second mechanism imparts primary significance to corrosion under pressure (ATKINSON, 1980; SWANSON, 1980). As recently established (MIRZOEV and NEGMATULLAEV, 1987), the filling of the basin, the subsequent operation of the turbines, and the periodical water discharges cause the increase of low seismicity and concurrent decrease of the number of stronger

[1] Water Economy Committee of Russian Federation, Moscow, Russia.
[2] Institute of Physics of the Earth, Russian Academy of Sciences, Moscow, Russia.

earthquakes. This effect is attributed to the reciprocal action of the intrapore pressure and the mechanical vibrations, which relieve the critical tangent stresses on seismoactive faults. In this process, even the microseisms contribute to stress release by an increase in the share of plastic deformations (MIRZOEV et al., 1991). The laboratory experiments (SOBOLEV et al., 1991) testify that the effect of impulsive mechanical oscillations on the contact of rocks may reduce by several tens of percent the threshold of critical tangent stresses causing the "stick-slip" phenomenon. One of the possible causes is the reduction of the normal to contact compressional stresses provoking the beginning of the unstable displacement. In the present paper, we discussed yet another mechanism causing earthquakes associated with the fluid-steam phase transition effected by heat flow (PANFILOV, 1990, 1994).

Description of the Model

Let us presume that at different depths in the crust two water reservoirs are situated one over the other. The first is near the surface, the second is below and filled with more heated water. We shall discuss schematically three cases (Fig. 1). If the exchange between the reservoirs is unimpeded (Fig. 1a) and the heating of the lower reservoir reaches the boiling point, then a free convection occurs and is maintained. If the reservoirs are joined by at least two water conducting channels,

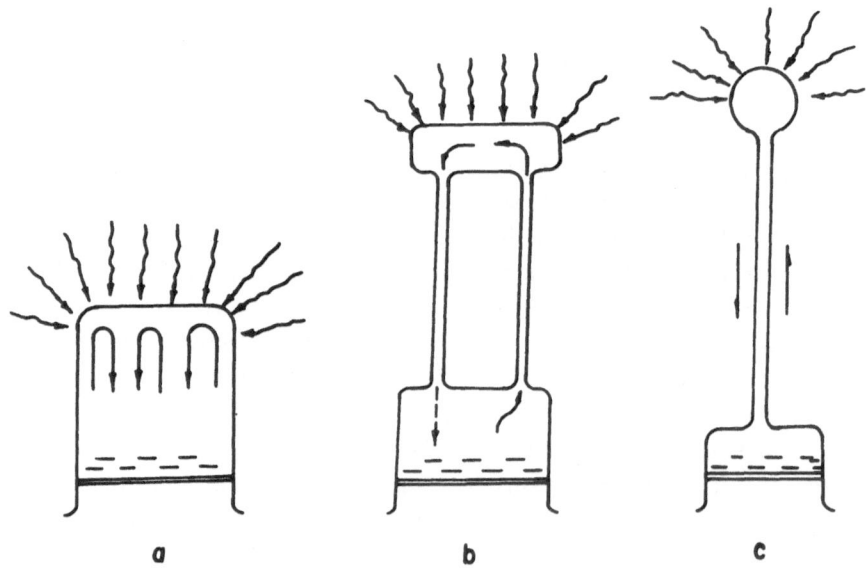

Figure 1
A principle scheme of steam-water convection; a — free convection, b — stable circulation; c — unstable convection.

then the water circulation is stable (Fig. 1b). If the lower and upper reservoirs are joined by one narrow channel (Fig. 1c), then the process alters essentially. In this case, there is no convection of fluid due to heating in the lower reservoir, and the water in the upper reservoir remains cold. When the heat flow is sufficiently intensive, the fluid in the lower reservoir is heated to a boiling point, which is conditioned by the properties of the fluid and by P-T specifications. At the moment of boiling of the fluid in the lower reservoir, the steam bubbles enter the narrow channel and reduce the pressure P in it due to the lesser volumetric weight of the water-steam mixture in the channel. This process releases the pressure in the lower reservoir and causes an avalanche-like boiling up of water in it. As a consequence, the fluid is dynamically ejected into the upper reservoir. The pressure and temperature in the lower reservoir fall, and as soon as the heat reserves in the lower reservoir are exhausted, the ejection ceases and a new portion of water is sucked in from the upper reservoir and, possibly, from the rocks surrounding the lower reservoir. The intervals between ejections are primarily controlled by the heat flow intensity and by the volume of the lower reservoir.

This process was simulated in the laboratory. Figure 2 shows the scheme of the device, composed of a lower retort of 720 cm^3, a larger upper container of 3700 cm^3, and the connecting tube 75 cm long and 0.8 cm in diameter. The total height of the water column in the system was 121 cm. The water in the lower retort was heated to the boiling point by an electric heater of 0.7 KW. The temperature and the pressure of the water in the lower retort were measured by sensors fixed at its middle; their measurements were recorded with an oscillograph. The frequency range of the recording sensor-oscillograph system varied from 0 to 400 Hz. The temperature of the water in the upper container was also controlled.

The unstable steam-water convection was studied with this model device as follows. The time interval of the first heating of the water in the lower retort from room temperature to the boiling point was 15–20 min. This procedure was followed by an intensive ejection of steam into the upper container and subsequent sucking of the water in the lower retort. From this moment the process continued by cycles, each of which contained three consecutive stages: heating, t_1; boiling up of water, t_2; ejection, t_3. The length of cycles shortened from t_1 to t_3. The structure of the cycle is shown by the scheme in Figure 3c.

The lower intensity of heating caused essential lengthening of the cycle by greater t_1 and t_2 periods. The duration of t_3 stage was practically stable. Visual observation showed that as soon as the first steam bubbles entered the connecting tube, the boiling process in the lower retort was sharply accelerated, and the ejection stage proceeded by separate surge with acoustic effects. The dynamics of the stage t_2 and t_3 can be traced by the records of sensors of pressure and temperature fixed on the lower retort (see Figs. 3a,b). According to visual observations, the high-frequency noise on the oscillogram of pressure is caused by the formation of individual bubbles of steam near the membrane of the sensor of pressure.

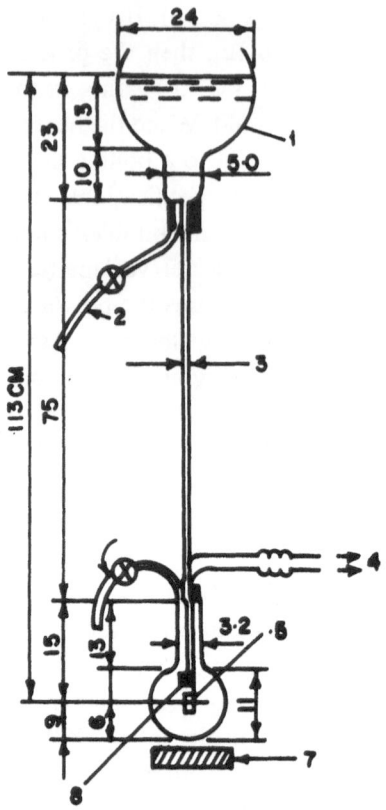

Figure 2

The scheme of the laboratory installation for the study of unstable steam-water convection. *Key*: 1 — the volume is 3700 cm³; 2 — discharge; 3 — the inner diameter is 8 mm; 4 — recording by oscillatograph; 5 — sensor of pressure; 6 — the volume; is 720 cm³ 3; 7 — heater; 8 — sensor of temperature.

In our case, the value of the change of pressure is represented best by the height of the column of fluid. At stage t_3, the maximal value of the fall of pressure was 120 cm; i.e., the pressure excessive to the atmospheric pressure, caused by the presence of the column of fluid in the system, was practically relieved.

In the lower retort, three stages were detected in the changes of water temperature. The first stage is connected with the lowering of temperature by about 1.5°C during the boiling up and ejection. The second and more significant lowering of temperature (by 5°C) is caused by the sucking in of cold water from the upper container. The subsequent slow rise of temperature indicated the beginning of the next cycle. The control of temperature in the upper container showed that it changes only after ejections. For example, as a result of 13 cycles the temperature of water in the upper container rose from 25 to 45°C.

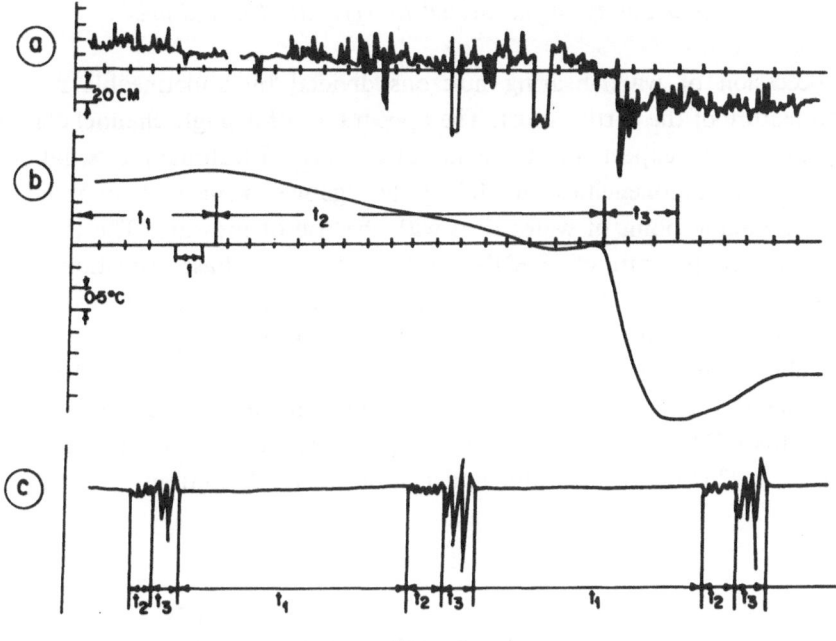

Figure 3
Changes in (a) pressure and (b) temperature in the lower retort (see Fig. 2) during development of unstable steam-water convection; (c) — scheme of cyclicity of the process.

In order to study the scale factor of the studied phenomenon, an installation, 4 m high was constructed with a 3 m long water column. The lower container was a metallic cylinder 400 mm in diameter, which included an aperture for visual observations. The water in this container was heated by inner heaters totalling 4.2 KW. The principal difference of this installation from the one described above was the mobile wall of the lower container which could be moved to the length of 50 mm. This wall was kept in the intermediate position by a spring and in the extreme positions by props, when the system was full or empty.

The process of boiling and ejection was accompanied by the visible reduction of pressure in the lower container with the loaded spring completely released for the entire 50 mm run to zero. The process of unstable steam-water convection in this installation was similar to the one described above, but the instability occurred when the water in the lower container was heated to 107°C. The system of signals showing the changes in the pressure allowed us to predict the time of dynamic ejections. It was established that the dynamic ejection in the system can be incited by lowering the pressure in the tube, or by injection of an air bubble into the lower container. The instability was completely removed by switching in a parallel tube between the upper and lower containers and inducing the fluid to circulate.

Applicability of the Model to Terrestrial Conditions

A succession of water-bearing horizons divided by impermeable rocks is a common feature of the earth's crust. The appearance of a single channel connecting the horizons can be expected in the zones of subvertical tectonic faults, taking into account their inhomogeneities and filling the "gauge" zone with material. As is known, the boiling point of water rises with the rise of pressure. The temperature of 374°C and the pressure of 23 MPa correspond to the phase transition when the density of water and of steam is equal, i.e, 0.33 g/cc. Such conditions can exist at a depth of a few kilometers from the earth's surface. The salinity of water raises the point of phase transition. Taking into account the lower density and salinity of the mixture, the *P-T* conditions of the phase transition fluid-steam may correspond to the depths of 5–7 km (see scheme in Fig. 4). The energy of the phase transition, which is 2257 kJ/kg for water under atmospheric conditions, at that point is zero.

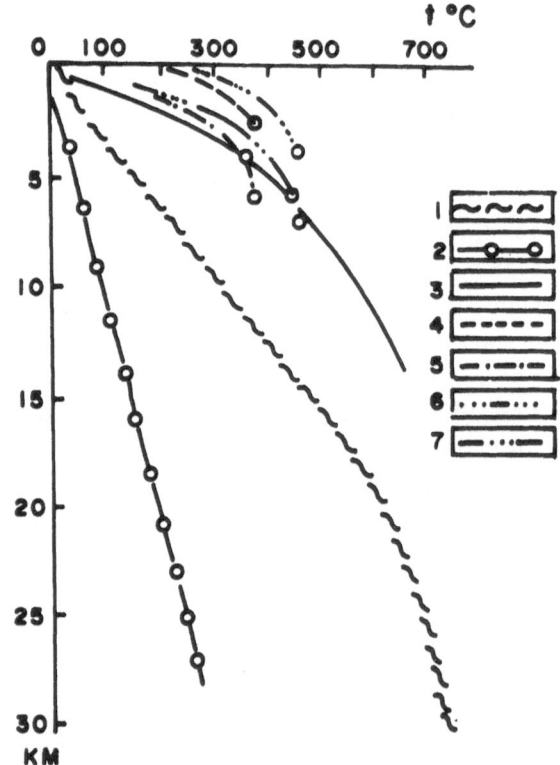

Figure 4
Distribution of temperature T with depth at the average (1), decreased (2) and (3) increased (3) gradients; the boiling temperature of water-steam transition at volumetric weight 1 g/cm³ 3 (4), at lesser volumetric weight of water due to heating (5), and for brines (6, 7).

It is important to note that the phase transition, while reducing pressure at its initiation point, encourages the propagation of the process by depth. Thus, the instability process. starting at a depth of several kilometers, may reach deeper, water-bearing horizons. It is difficult to estimate quantitatively the lower boundary of the phenomenon at the present stage of research.

The unstable steam-water convection can trigger an earthquake, as a result of two major mechanisms. A sudden change in pressure inside a water-saturated tectonic fault under stress may initiate instability by reducing the friction coefficient and subsequent displacement of the sides of the fault. Moreover, the dynamically occurring ejection effect excites the mechanical oscillations, as shown above. They can form a triggering mechanism of displacement in the neighboring fault by a temporary reduction of normal or an increase of tangent stresses.

The Gazli earthquake is one example of this model. The event is peculiar because three strong earthquakes ($M > 7.0$) occurred in quick succession, and their foci were located at distances comparable with the size of the foci themselves. Figure 5 is a schematic representation of this event, which shows the stratum gas pressure, caused by the pumping out of gas, correlated with the temperature of the water-steam phase transition (PISKULIN and RAIZMAN, 1985). The thermodynamic pressure-temperature parameters, under the exploitation conditions of the Gazli deposit, admit the appearance of the described trigger mechanism as a result of the unstable steam-water convection.

Conclusion

Owing to the level of our knowledge regarding the properties of deeply deposited rocks in the earth's crust, the suggested model of initiation of earthquakes is hypothetical. On its basis however, certain effects accompanying the earthquakes can be explained:

(a) A sharp rise of the level and temperature of the upper water-bearing horizons observed, for example, during the Tashkent earthquake in 1966;

(b) The spouting of wells and bore-holes, which occurred during the catastrophic earthquakes in Gazli in 1984 and Tangshan in 1976;

(c) The underground booming noise preceding the seismic shock;

(d) The "floating up" of hypocenters of aftershocks and greater time intervals between later aftershocks.

The last point needs further explanation. As the rocks surrounding the focus cool off, the unstable steam-water convection (the trigger of the aftershcok) shall exist at still lower pressure, i.e., at a lesser depth. The intensity of the heat flow to the focus from the surrounding rocks also diminishes, causing longer intervals in the operation of the trigger.

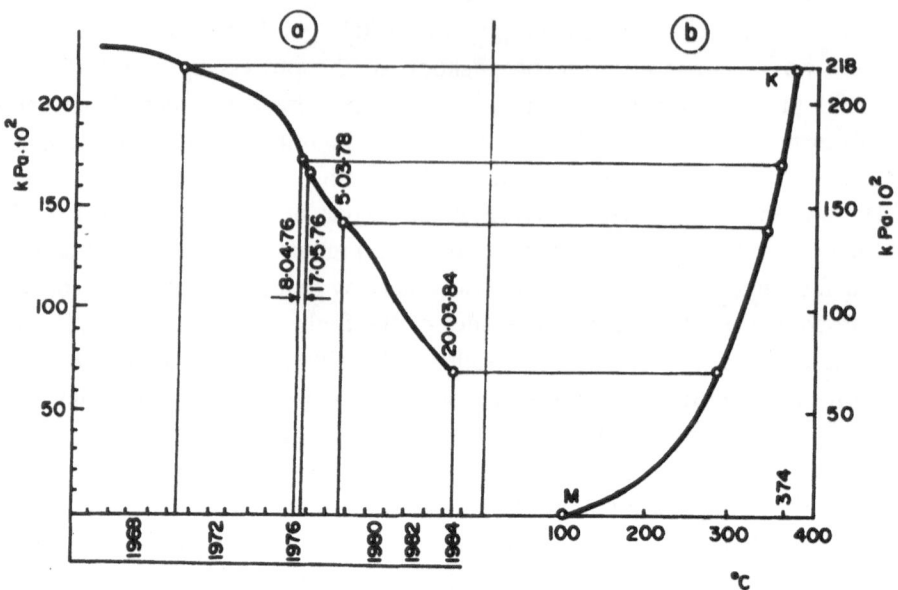

Figure 5
Correlation between gas pressure (a) and water-steam transition (b) as applied to Gazli conditions.

Lastly, the phenomenon of the excited seismicity, while filling the water storage basin, can partly be attributed to the appearance of new sources for the trigger mechanism by water-saturation of the nearest deep faults.

Acknowledgements

The authors extend their thanks to V. A. Kalinin, A. O. Gliko and other colleagues at the Institute of Physics of the Earth, RAS, for the discussion of the ideas proposed in this paper and for their useful comments.

REFERENCES

ATKINSON, B. K. (1980), *A Fracture Mechanism Study of Subcritical Tensible Cracking of Quartz in Wet Environment*, Pure and Appl. Geophys. *117*, 1011–1024.

GUPTA, H. K., *Reservoir-induced Earthquakes* (Elsevier, Amsterdam-London-New York-Tokyo 1992). 355 pp.

HUBERT, H. K., and RUBEY, W. W. (1959), *Role of Fluid Pressure in Mechanics of Overthrust Faulting*, Bull. Geol. Soc. Am., 115–166.

MIRZOEV, K. M., and NEGMATULLAEV, S. KH. (1987), *The Influence of Mechanical Vibrations on Release of Seismic Energy*, Earthquake Prediction. Dushanbe-Moscow, Donish, 365–372.

MIRZOEV, K. M., VINOGRADOV, S. D., and RUZIBAEV, Z. (1991). *Effect of Microseisms and Vibration on Acoustic Emission*, Izvestiya AN SSSR, Fizika Zemli *12*, 69–72.

PAKISER, L. C., EATON, J. P., HEALY, J. H., and RALEIGH, C. B. (1969), *Earthquake Prediction and Control*, Science *166*, 1467–1479.

PANFILOV, V. S. (1990), *Hydrogeothermal Theory of Earthquakes—Basis of Prediction Warning Systems, and Counteractions. Hydrotechnical Construction* (translated from Russian), Consultants Bureau, February, August, *24*, 142–148.

PANFILOV, V. S. (1994), *Hydrogeothermal Phenomena in Seismic Genesis*, Physics of the Earth *2*, 79–87.

PISKULIN, V. A. and RAIZMAN, A. P. (1985), *Deformations of the Earth's Surface in the Gazli Region*, Geodeziya i Kartografiya *9*, 53–57.

SIMPSON, D. W. (1986), *Triggered Earthquakes*, Ann. Rev. Earth Planet. Sci. *14*, 21–42.

SOBOLEV, G. A., KOL'TSOV, A. V., and ANDREEV, V. O. (1991), *The Trigger Effect of Oscillations in the Earthquakes Model*, Doklady AN SSSR *319*, 337–341.

SWANSON, P. L. (1980), *Stress Corrosion Cracking in Westerly Granite: An Examination by the Double Torsion Technique*, H. Sc. Thesis. Univ. Colorado, U.S.A., 142 pp.

(Received May 27, 1994, revised/accepted November 3, 1994)

PAGEOPH, Vol. 145, No. 1 (1995)

0033–4553/95/010149–05$1.50 + 0.20/0

Assessment of Potential Strength of an Induced Earthquake by Using Fuzzy Multifactorial Evaluation

DEYI FENG,[1] XUEJUN YU,[2] and JINQPING GU[3]

Abstract—The potential strength of an induced earthquake depends on a series of factors at different levels. It can be effectively assessed and predicted by using the method of fuzzy multifactorial evaluation from a fuzzy set theory. As an illustration of the above-mentioned method, this paper has applied the method to assessing the potential strength of induced earthquakes due to water reservoir.

Key words: Potential strength of earthquakes, fuzzy set theory.

Introduction

As is well-known, the assessment of the potential strength of the induced earthquake in a reservoir region, gas field or mine region, is a very important problem for many countries, because the relatively strong induced seismic event can cause serious damage and losses (GUHA, 1990; GUPTA, 1990; PLOTNIKOVA *et al.*, 1990). At the same time, it is a very difficult and complex problem, because the potential strength of the induced earthquake depends on a series of factors in the different ranges (HU, 1992; SIMPSON, 1990; RAJENDRAN and TALWANI, 1992). They include the geological settings, tectonic conditions, rock types and properties, geomorphic condition and others. Several methods have been applied to the hazard assessment and prediction of the induced earthquake (SKIPP, 1981; CHANG, 1990; BEACHER and KEENEY, 1982; FENG *et al.*, 1984/85). Among them we can mention with emphasis the method of fuzzy multifactorial evaluation from a fuzzy set theory (ZADEH, 1965) which may be used effectively in this type of problem.

Illustrating the above-mentioned method, this paper has applied the two-level fuzzy multifactorial evaluation to assess the potential strength of the induced earthquake due to the water reservoir.

---

[1] Seismological Bureau of Tianjin, Tianjin 300201, China.
[2] Seismological Bureau of Zhejiang, Hangzhou 310013, China.
[3] Seismological Bureau of Jiangsu, Nanjing 210014, China.

Method

The method of multi-level fuzzy mutlifactorial evaluation was advanced by WANG PEIZHUANG (1983). Its main steps are as follows:

1. Choosing the influence factors and dividing them into different levels:

$$U = \{U_1, U_2, U_3, U_4\}.$$

The chosen influence factors were divided into two levels, namely, high levels U_1, U_2, U_3, U_4 and low levels e_1, e_2, e_3; s_1, s_2, s_3, etc. They are: (1) Synthetic influence parameters $U_1 = \{e_1, e_2, e_3\}$, where $e_1 < 2.20$, $2.20 \leqslant e_2 < 3.46$, $e_3 \geqslant 3.46$; (2) Geological settings $U_2 = \{s_1, s_2, s_3\}$, where s_1, s_2, s_3 denote the thrust slip fault, normal fault and strike-slip fault, respectively; (3) Tectonic conditions $U_3 = \{f_1, f_2\}$, where f_1, f_2 denote the active and nonactive faults; (4) Rock types $U_4 = \{g_1, g_2, g_3, g_4\}$, where g_1, g_2, g_3, g_4 denote the sedimentary, metamorphic, magmatic and carbonatite rocks, respectively.

2. Establishing the set of assessment: The potential strength of a maximum induced earthquake in a given region was classified into 3 ranges of magnitudes: $M_1 < 3.5$, $M_2 = 3.6 - 4.9$, $M_3 \geqslant 5.0$. They form a set of assessment $H = \{H_1, H_2, H_3\}$.

3. Determining the sets of weights of influence factors at two levels:

$$A(A_1, A_2, A_3, A_4),$$

where $A_1 = (A_1(H_1), A_1(H_2), A_1(H_3))$, $A_1(H_i) = (A_{e1i}, A_{e2i}, A_{e3i})$, $i = 1, 2, 3$, etc. They can be determined statistically or empirically.

The sets of weights of influence factors at two levels determined empirically are as follows (FENG, 1992):

$$\mathbf{A} = (A_1, A_2, A_3, A_4)$$

$$= (0.30, 0.25, 0.20, 0.25)$$

$$\mathbf{A_e} = \begin{array}{c} \\ \\ \\ \\ \end{array} \begin{array}{ccc} e_1 & e_2 & e_3 \\ \hline 0.000 & 0.900 & 0.100 \\ 0.250 & 0.500 & 0.250 \\ 0.000 & 0.500 & 0.500 \end{array} \begin{array}{c} \\ H_1 \\ H_2 \\ H_3 \end{array}$$

$$\mathbf{A_s} = \begin{array}{c} \\ \\ \\ \\ \end{array} \begin{array}{ccc} s_1 & s_2 & s_3 \\ \hline 0.313 & 0.438 & 0.250 \\ 0.200 & 0.600 & 0.200 \\ 0.100 & 0.500 & 0.400 \end{array} \begin{array}{c} \\ H_1 \\ H_2 \\ H_3 \end{array}$$

$$\mathbf{A_f} = \begin{array}{c} \\ \\ \\ \\ \end{array} \begin{array}{cc} f_1 & f_2 \\ \hline 0.333 & 0.667 \\ 1.000 & 0.000 \\ 0.800 & 0.200 \end{array} \begin{array}{c} \\ H_1 \\ H_2 \\ H_3 \end{array}$$

$$\begin{array}{cccc}
g_1 & g_2 & g_3 & g_4
\end{array}$$

$$\mathbf{A_g} = \begin{vmatrix}
0.263 & 0.263 & 0.105 & 0.369 \\
0.231 & 0.269 & 0.077 & 0.423 \\
0.308 & 0.154 & 0.231 & 0.308
\end{vmatrix} \begin{matrix} H_1 \\ H_2 \\ H_3 \end{matrix}$$

$e_1 < 2.20$, $2.20 \leqslant e_2 < 3.46$, $e_3 \geqslant 3.46$;

s_1, s_2, s_3: geological settings;

f_1, f_2: tectonic conditions;

g_1, g_2, g_3, g_4: types of rocks.

4. Determining the relationship matrices of influence factors:

$$R_e = \{r_{ij}\}, \ R_s = \{r'_{ij}\}, \ R_f = \{r''_{ik}\}, \ R_g = \{r''_{im}\}; \ i = j = 1, 2, 3; \ k = 1, 2; \ m = 1, 2, 3, 4.$$

The relationship matrices of influence factors determined on the basis of observational data are as follows (FENG, 1992):

$$\begin{array}{ccc}
H_1 & H_2 & H_3
\end{array}$$

$$\mathbf{R_e} = \begin{vmatrix}
0.000 & 1.000 & 0.000 \\
0.450 & 0.300 & 0.250 \\
0.273 & 0.273 & 0.454
\end{vmatrix} \begin{matrix} e_1 \\ e_2 \\ e_3 \end{matrix}$$

$$\begin{array}{ccc}
H_1 & H_2 & H_3
\end{array}$$

$$\mathbf{R_s} = \begin{vmatrix}
0.625 & 0.250 & 0.125 \\
0.389 & 0.333 & 0.278 \\
0.400 & 0.200 & 0.400
\end{vmatrix} \begin{matrix} s_1 \\ s_2 \\ s_3 \end{matrix}$$

$$\begin{array}{ccc}
H_1 & H_2 & H_3
\end{array}$$

$$\mathbf{R_f} = \begin{vmatrix}
0.231 & 0.461 & 0.308 \\
0.857 & 0.000 & 0.143
\end{vmatrix} \begin{matrix} f_1 \\ f_2 \end{matrix}$$

$$\begin{array}{ccc}
H_1 & H_2 & H_3
\end{array}$$

$$\mathbf{R_g} = \begin{vmatrix}
0.313 & 0.375 & 0.313 \\
0.357 & 0.500 & 0.143 \\
0.286 & 0.286 & 0.428 \\
0.333 & 0.524 & 0.143
\end{vmatrix} \begin{matrix} g_1 \\ g_2 \\ g_3 \\ g_4 \end{matrix}$$

5. Two-level fuzzy mutlifactorial evaluation of the potential strength of the induced earthquake.

The formula for two-level fuzzy multifactorial evaluation of the potential strength of the induced earthquake can be represented as:

$$\mathbf{B} = b_1/H_1 + b_2/H_2 + b_3/H_3 = \mathbf{A} \cdot \mathbf{R}$$

$$= \mathbf{A} \cdot \begin{bmatrix}
\mathbf{A_e} \cdot \mathbf{R_e} \\
\mathbf{A_s} \cdot \mathbf{R_s} \\
\mathbf{A_f} \cdot \mathbf{R_f} \\
\mathbf{A_g} \cdot \mathbf{R_g}
\end{bmatrix}$$

where symbol " · " denotes the compositional operation on matrices and we have,

$$\mathbf{B} = \mathbf{A} \cdot \mathbf{R} = [b_j]$$

$$b_j = \max_k \min(a_k, r_{kj}); \quad j = 1, 2, 3.$$

Here **B** is the fuzzy set for the assessment result, and the optimum assessment can be obtained according to the maximum component of **B**, i.e.,

$$b_{\max} = \max[b_1, b_2, b_3];$$

b_1, b_2 and b_3 are the membership degrees of occurrence of the earthquake with a magnitude in H_1, H_2 and H_3 ranges, respectively. Actually, we can assess the potential strength (magnitude) of the induced earthquake more accurately by using all three components b_1, b_2 and b_3.

Results

By using the suggested method, the potential strengths of 15 induced earthquakes due to ·water reservoirs in Canada, China, India, Greece, U.S.A. and Zambia

Table 1

The results of assessment of potential strength · of induced earthquakes due to water reservoirs using two-level fuzzy multifactorial evaluation

Earthquake due to reservoir	Magnitude	Chosen influence factors	Set of assessment	Result of assessment	Result of assessment by E only
Nanshui (China)	2.4	e_2, s_2, s_3, f_2, g_4	(0.534 0.287 0.179)	2.5	3.9
Jocassee (U.S.A.)	2.5	e_2, s_2, f_2, g_2	(0.546 0.299 0.155)	2.5	3.9
Longyangxiya (China)	3.2	e_2, s_1, s_3, f_1, g_2	(0.375 0.392 0.233)	3.5–4.2	3.8
Wuchijiang (China)	2.8	e_2, s_1, s_3, f_1, g_3	(0.380 0.261 0.359)	2.8	3.9
Toling (China)	3.2	e_2, s_2, s_3, f_1, g_4	(0.363 0.405 0.232)	3.5–4.2	4.6
Manicouagan 3 (Canada)	3.8	e_1, s_1, f_1, g_2	(0.303 0.600 0.097)	3.5–4.2	3.2
Fouzhiling (China)	4.5	e_3, s_2, f_1, g_2, g_4	(0.215 0.421 0.364)	4.3–4.9	4.8
Danjiangkou (China)	4.7	$e_3, s_1, s_3, f_1, g_1, g_4$	(0.255 0.336 0.409)	5.0–5.7	5.1
Shengwuo (China)	4.8	$e_3, s_2, f_1, g_1, g_2, g_3, g_4$	(0.206 0.388 0.406)	5.0–5.7	4.7
Oroville (U.S.A.)	5.4	e_2, s_2, f_1, g_3	(0.338 0.318 0.350)	5.9	4.2
Marathon (Greece)	5.75	e_3, s_2, f_1, g_2, g_3	(0.201 0.376 0.417)	5.0–5.7	4.8
Xinfengjiang (China)	6.1	e_2, s_3, f_1, g_3	(0.331 0.254 0.415)	6.0	4.6
Kariba (Zambia)	6.1	$e_3, s_2, f_2, g_1, g_2, g_3$	(0.374 0.249 0.376)	5.9	5.9
Kremasta (Greece)	6.2	e_3, s_2, f_1, g_1, g_4	(0.206 0.394 0.400)	5.0–5.7	5.8
Koyna (India)	6.5	e_3, s_3, f_1, g_3	(0.178 0.251 0.571)	6.5	5.4

with magnitudes 2.0–6.5 were assessed and examined. The results obtained are shown in Table 1. For comparison, the result of assessment by using only one parameter E (comprehensive effecting coefficient; CHANG, 1990) is also shown in the table. From Table 1 we can see that the method of fuzzy multifactorial evaluation, especially the multi-level evaluation method, can be used to assess the potential strength of the induced earthquake more effectively and more accurately.

REFERENCES

BEACHER, G. B., and KEENEY, R. L. (1982), *Statistical Examination of Reservoir-induced Seismicity*, Bull. Seismol. Soc. Am. *92*, 552–569.

CHANG BAOQI (1990), *Preliminary Study on the Prediction of Reservoir Earthquakes*, Gerland. Beitr. Geophysik *99*, 407–424.

FENG DEYI, JING-PING GU, MING-ZHOU-LIN, SHAO-XIE XU, and XUE-JUN YU. (1984/85), *Assessment of Earthquake Hazard by Simultaneous Use of the Statistical Method and the Method of Fuzzy Mathematics*, Pure and Appl. Geophys. *122*, 982–987.

FENG DEYI. *Fuzzy Seismology* (Seismological Press, Beijing 1992) *14*, pp. 122–126.

GUHA, S. K. (1990), *Large Water-reservoir Related Induced Seismicity*, Gerland. Beitr. Geophysik *99*, 265–288.

GUPTA, H. K. (1990), *Artificial Water Reservoirs and Earthquakes; A Worldwide Status*, Gerland. Beitr. Geophysik *99*, 221–228.

HU, P. (1992), *Advances in Reservoir-induced Seismicity Research in China*, Tectonophys. *209*, 331–337.

RAJENDRAN, K., and TALWANI, P. (1992). *The Role of Elastic, Undrained and Drained Responses in Triggering Earthquakes at Monticello Reservoir, South Carolina*, Bull. Seismol. Soc. Am. *82*, 1867–1888.

PLOTNIKOVA, L. M., FLYONOVA, M. G., and MACHMUDOVA, V. I. (1990), *Induced Seismicity in the Gazli Gas Field Region*, Gerland. Beitr. Geophysik. *99*, 389–399.

SIMPSON, D. W. (1990), *Inhomogeneities in Rock Properties and their Influence on Reservoir-induced Seismicity*, Gerland. Beitr. Geophysik *99*, 205–219.

SKIPP, S. K., *The Potential for Induced Seismicity: Geological Approaches, Dams and Earthquakes* (Thomas Telford Ltd., London 1981) pp. 297–296.

WANG PEIZHUANG, *Fuzzy Sets Theory and its Applications* (Shanghai Scientific and Technical Publishers, Shanghai 1983) pp. 91–94.

ZADEH, L. A. (1965), *Fuzzy Sets*, Inform. Control *8*, 338.

(Received April 8, 1994, revised/accepted October 18, 1994)

PAGEOPH, Vol. 145, No. 1 (1995)

0033–4553/95/010155–11$1.50 + 0.20/0

Role of Dykes in Induced Seismicity at Bhatsa Reservoir, Maharashtra, India

R. K. Chadha[1]

Abstract — The geological and hydrological conditions near the reservoir site play an important role in the generation or absence of seismic activity. Near Bhatsa reservoir, along the west coast of India intense seismic activity occurred during August–September 1983, after a lag of six years of initial impounding. From July 1983 to September 1990, 15,388 earthquakes (mostly $M_1 < 3.0$) were recorded, the largest being of magnitude 4.9. The spatial distribution of well located 172 earthquakes suggest a strong correlation between the epicenters and the disposition of dykes and faults around the Bhatsa region. It is inferred that these dykes have acted as "barriers" for the diffusion of water from the reservoir, thereby becoming zones of instability due to increased pore pressure not only along them but also over the volume they bound.

Key words: Dykes, pore pressure.

Introduction

The Bhatsa region in Maharashtra, India experienced intense seismic activity in the latter half of 1983. It is situated 200 km north of the famous reservoir-induced seismicity (RIS) site at Koyna along the west coast of India (Fig. 1). The construction of the dam (height 88.5 m, when completed) began in 1968 and the first impounding reaching 17 m was done in 1977. Subsequently, the dam height was raised in stages, every year. In July, 1983 the lake level was raised by 18 m in a month. This rapid increase of the water level was followed by a burst of seismic activity in August and September 1983, with significant events of M_L magnitudes 4.4 and 4.9 on August 17 and September 15, 1983, respectively. Earlier studies of these earthquakes (RASTOGI *et al.*, 1986; CHADHA, 1991) have correlated initial seismicity with water levels, focal mechanism, growth of epicentral area, foreshock-aftershock pattern, *b* values and M_1/M_0 ratio to categorize this activity as

---

[1] National Geophysical Research Institute, Hyderabad - 500 007, India.

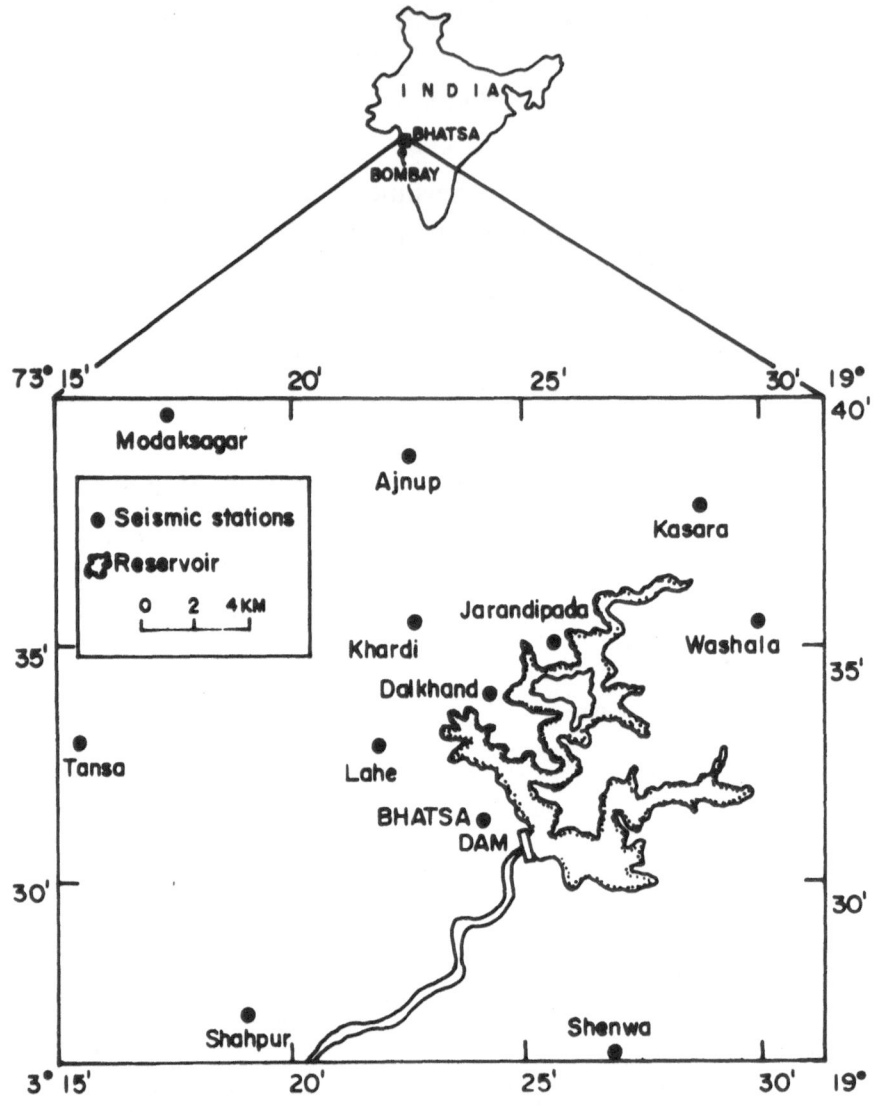

Figure 1
Location of Bhatsa and network of seismic station around the reservoir.

induced seismicity as described by GUPTA and RASTOGI (1976). In this study, the earthquakes which occurred from October 1983 to April 1985 are relocated along with earthquakes during May 1985 to September 1990. An attempt is made here to correlate the earthquakes epicenters with dykes and faults in the Bhatsa region and to introduce the role played by these geological structures in determining seismicity. The study also tries to provide a mechanism for these induced earthquakes.

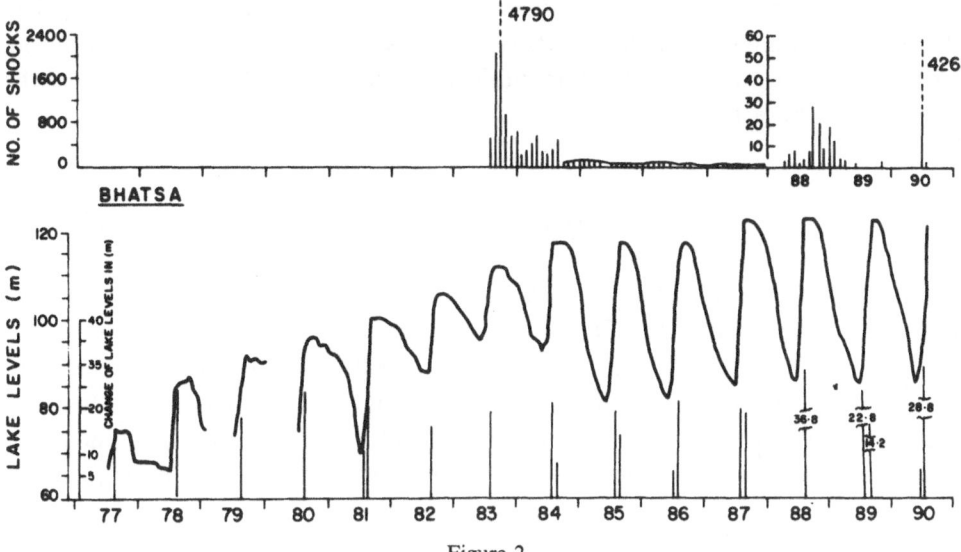

Figure 2

Water levels and changes of lake level of Bhatsa reservoir from 1977 to 1990. Monthly frequency of earthquakes is also shown.

Seismicity and Reservoir Impounding

The Bhatsa reservoir lies within the moderately active western coastal margin and falls in zone III of the Seismic Zoning Map of India (ISI 1893–1975). The west coast of India has historically been prone to mild earthquakes (BAPAT *et al.*, 1983). However, no earthquake has been reported from the Bhatsa reservoir area. The first impounding of the reservoir upto 75 m (AMSL, 17 m above the river bed) was done during 1977. The dam was subsequently raised in stages every year. Figure 2 shows the water level and the seismicity from 1977 to 1990. During the month of July 1983, the reservoir water level was raised to 114 m. The water level rose about 18 m in July 1983 alone. Although, the first tremor in the region was felt in May 1983, the rapid loading in July 1983 was followed by intense seismic activity during July–October 1983. The peak seismicity occurred in September 1983 and included the largest earthquake of magnitude 4.9 of September 15, 1983. The seismic activity started declining from October 1983 onwards. Figure 2 shows that the earthquakes started occurring in the Bhatsa reservoir area when the maximum lake level was raised for the first time to 110 m in 1983. Later in 1984 the lake level was further raised to 115 m but the declining trend of the seismicity continued except during August 1984, when there was a slight increase in the seismicity following the rainy season. The maximum lake level was maintained at 115 m through 1986. In 1987 the maximum lake level was further increased to 120 m and this maximum level was achieved every year until 1990. In June 1990 when the reservoir level was rising following the rainfall, a spurt of seismicity occurred with a felt earthquake. A spurt of seismicity with 692 earthquakes, mostly less than 3.0 magnitude occurred during June 1990 but died down within a month's time. Since then, the seismicity near the Bhatsa reservoir is low.

Geology and Tectonics

The Bhatsa reservoir area lies in the Deccan volcanic province which covers one third of the Peninsular India. The Deccan lava erupted in the late Mesozoic-Cenozoic period (60–65 my) covering an area of 500,000 sq km. The total thickness of Deccan lava is estimated to vary between 1 to 2 km. The Bhatsa region lies in a 5 km wide NW trending belt of high density lineaments (Fig. 3) the boundary of which is marked by the Kengrinadi lineament in the north and Kalu-Surya in the

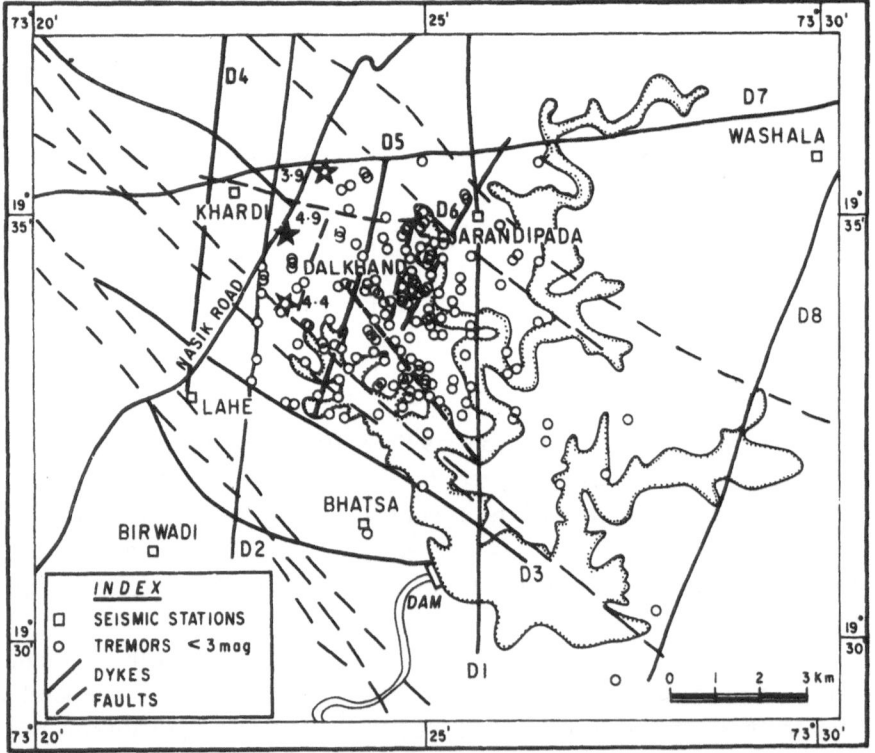

Figure 3
Epicenters of earthquakes located from October 1983 to June 1990 along with faults/fractures and dykes in the region. The faults and dykes shown in the map are taken from GSI and Poona University reports. The foreshock, main shock and the largest aftershock are denoted by stars with their respective magnitudes.

south (PATIL *et al.*, 1986). Both of these lineaments can be traced for several hundred kilometers. A series of parallel NW–SE trending faults that exhibit vertical and lateral movements and create blocks of horst and graben like structure in the region, have been mapped by the Geological Survey of India (1984). These faults are intersected by numerous dolerite and basaltic dykes for distances of 5–15 km. The oldest prominent direction of these dykes are E–W followed in order NE–SW and N–S.

Data and Analysis

The seismicity around the Bhatsa reservoir, which started occurring in July, 1983 was recorded by 12 seismic stations network operated by the National Geophysical Research Institute (NGRI), Hyderabad, India and Maharashtra Engineering Research Institute, Nasik. An analog telemetry network was established around Bhatsa by Bhaba Atomic Research Centre, Bombay in 1986. A total number of 15,388 events were recorded between July 1983 and September 1990. From these, 2 events were of magnitude > 4, 17 were between magnitude 3 and 4 and the remaining were < 3 (Table 1). Peak seismicity occurred during August and September 1983, when 2072 and 4790 earthquakes were recorded by the network, respectively. The activity started declining from October 1983 onwards. Earthquakes of magnitude > 2 frequently occurred until August 1984. Initial seismic activity from October 1983 to April 1985 was located by RASTOGI *et al.* (1986). Using the HYP071 computer program by LEE and LAHR (1972), earthquakes recorded at 4 or more seismic station were located for the period October 1983 to September 1990.

Only those events whose root mean square (rms) of *P* residual is <0.1 s, average horizontal error (erh) and error in depth (erz) is <1.0 km, are utilized in this study. The velocity model used for locating these earthquakes was obtained by Deep Seismic Sounding (DSS) from Koyna Profile II, 200 km south of the Bhatsa reservoir (KAILA *et al.*, 1981a,b).

There are 172 events located in the area of about 5 × 7 km between the Bhatsa Dam and Khardi town. Figure 3 shows the epicenters of these events along with faults/fractures and dykes in the region. This figure clearly illustrates that 90% of the epicenters are confined to an area limited by dykes D1, D2, D3 and D5. This block is cut-up by several other minor faults/fractures and dykes in the NW–SE, NE–SW and E–W directions, making it highly fractured and thus more porous. Only 10% of the epicenters are in the other half of the reservoir. The foreshock of magnitude 4.4 on August, 17, the main shock of magnitude 4.9 on September 15, 1983 and the largest aftershock on January 7, 1984 of magnitude 3.9, occurred within this block and are aligned in N–S direction close to Dyke 2 in the same direction. The composite fault plane solution plotted for the events within this

Table 1

Number of earthquakes recorded at Bhatsa seismic station

Month	No. of earthquakes		Magnitudes		
	Recorded	Felt	< 3	> 3 < 4	> 4
(1)	(2)	(3)	(4)	(5)	(6)
1983					
July	516	1	515	1	–
August	2072	25	2068	3	1
September	4790	215	4778	11	1
October	955	36	955	–	–
November	571	5	571	–	–
December	636	8	636	–	–
1984					
January	217	3	216	1	–
February	294	6	294	–	–
March	421	10	421	–	–
April	610	7	610	–	–
May	271	6	271	–	–
June	262	6	262	–	–
July	346	7	346	–	–
August	510	7	510	–	–
September	85	–	85	–	–
October	88	–	88	–	–
November	105	–	105	–	–
December	144	–	144	–	–
1985					
January	151	–	151	–	–
February	109	–	109	–	–
March	165	–	165	–	–
April	112	–	112	–	–
May	180	–	180	–	–
June	58	–	58	–	–
July	29	–	29	–	–
August	38	–	38	–	–
September	51	–	51	–	–
October	52	–	52	–	–
November	32	–	32	–	–
December	63	–	63	–	–
1986					
January	67	–	67	–	–
February	81	–	81	–	–
March	87	–	87	–	–
April	54	–	54	–	–
May	36	–	36	–	–
June	02	–	02	–	–
July	03	–	03	–	–
August	43	–	43	–	–
September	26	–	26	–	–

Table 1 (*Contd.*)

Month	No. of earthquakes		Magnitudes		
	Recorded	Felt	<3	$>3<4$	>4
(1)	(2)	(3)	(4)	(5)	(6)
1986					
October	11	–	11	–	–
November	15	–	15	–	–
December	24	–	24	–	–
1987					
January	42	–	42	–	–
February	34	–	34	–	–
March	11	–	11	–	–
April	02	–	02	–	–
May	05	–	05	–	–
June	–	–	–	–	–
July	10	–	10	–	–
August	06	–	06	–	–
September	–	–	–	–	–
October	01	–	01	–	–
November	05	–	05	–	–
December	04	–	04	–	–
1988					
January	–	–	–	–	–
February	–	–	–	–	–
March	03	–	03	–	–
April	03	–	03	–	–
May	06	–	06	–	–
June	03	–	03	–	–
July	–	–	–	–	–
August	11	–	11	–	–
September	29	–	29	–	–
October	26	–	26	–	–
November	10	–	10	–	–
December	20	–	20	–	–
1989					
January	12	–	12	–	–
February	04	–	04	–	–
March	03	–	03	–	–
April	–	–	–	–	–
May	02	–	02	–	–
June	–	–	–	–	–
July	–	–	–	–	–
August	–	–	–	–	–
September	–	–	–	–	–
October	03	–	03	–	–
November	–	–	–	–	–
December	01	–	01	–	–

Table 1 (*Contd.*)

Month	No. of earthquakes		Magnitudes		
	Recorded	Felt	< 3	> 3 < 4	> 4
(1)	(2)	(3)	(4)	(5)	(6)
1990					
January	–	–	–	–	–
February	–	–	–	–	–
March	–	–	–	–	–
April	–	–	–	–	–
May	–	–	–	–	–
June	692	10	691	1	–
July	04	–	04	–	–
August	08	–	08	–	–
September	46	–	46	–	–
Total	15388	352	15369	17	2

block, where first motions are very clear, indicates two nodal planes in NE–SW and NW–SE directions showing strike-slip movement (Fig. 4). The NE–SW nodal plane which shows left-lateral movement is preferred to be the fault plane as it

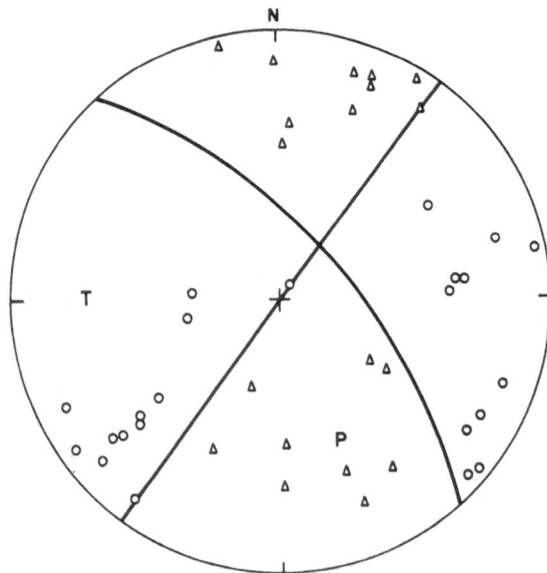

Figure 4
Composite fault plane solution of the earthquakes that occurred within the block limited by dykes D1, D2, D3 and D5. Two nearly vertical nodal planes are shown in NE–SW and NW–SE indicating strike-slip faulting. "*P*" and "*T*" are axes of maximum and minimum compression.

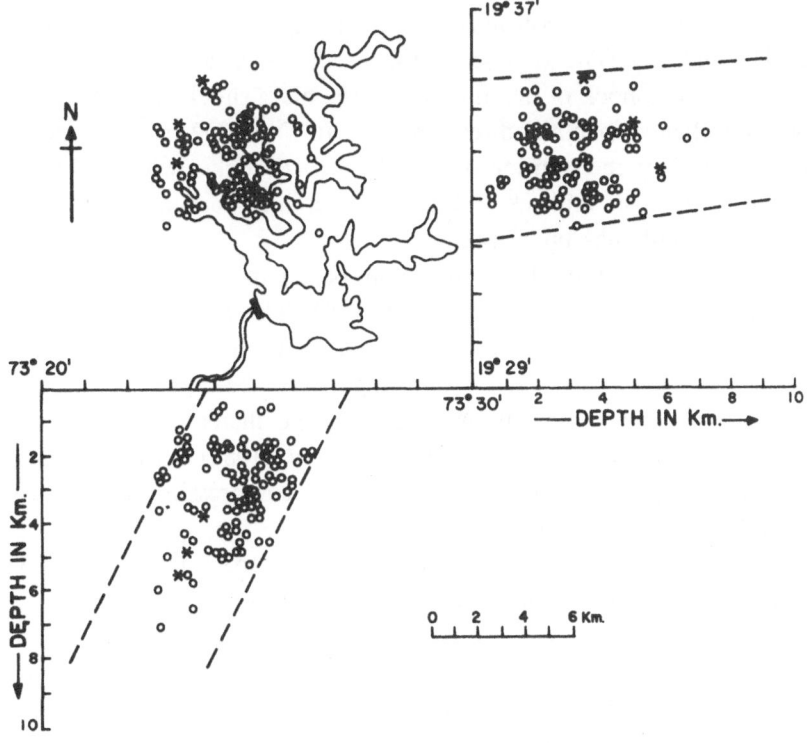

Figure 5
Projection of earthquakes foci on latitudinal and longitudinal planes. The hypocenters are confined up
to depth of 5 km.

coincides with the alignment of foreshock, main shock and the largest aftershock along the direction of Dyke D2.

Figure 5 shows the projection of earthquake foci on latitudinal and longitudinal planes. The depth of the earthquakes are mostly up to 5 km with a few reaching 7 km. The figure shows that along a N–S longitudinal plane, the earthquake foci are distributed along a steeply easterly dipping plane. The alignment of the three major earthquakes in a NNE–SSW direction, coinciding with the NE–SW nodal plane obtained from CFPS, indicate that the initial seismicity was triggered along the Dyke D2 in NNE direction.

Discussion

The spatial and temporal distributions of seismicity following the impounding of the Bhatsa reservoir Maharashtra appear to have activated a zone 5×7 km upstream of the Bhatsa reservoir north of the dam limited by four prominent dykes, D1, D2, D3 and D5. The depths of the earthquakes are mostly < 5 km. The epicenters of the foreshock of magnitude 4.4, the main shock of 4.9, and the largest aftershock are aligned in NNE direction close to Dyke D2 in the same direction. The composite fault plane solution obtained for earthquakes with the block defined by the four dykes also indicate a NE–SW trending vertical fault along which left-lateral movements has taken place.

The impounding of the Bhatsa reservoir began in 1977 and the first seismic activity in the region was noticed in May 1983. The time-lag of about 6 years between the initial loading and the onset of seismicity in the region suggests that it was a case of "delayed response" which might have been caused by diffusion of water along weak zones, this diffusion caused the increase in pore pressure at seismogenic depth and lead to failure. SIMPSON *et al.* (1988) described the rapid and delayed response depending upon the time scales in compression- and diffusion-controlled changes in pore pressure.

The rock beneath a reservoir and the structures surrounding it play a vital role in controlling the genesis of natural as well as induced earthquakes. In the Bhatsa region, the Deccan traps are highly fractured and faulted and therefore highly heterogeneous and permeable. A number of dykes criss-cross the reservoir area. In such an environment diffusion of water from the reservoir may lead to the formation of zones of stress concentrations along certain localized pockets where higher pore pressure can develop. In the Bhatsa reservoir area, the dykes D1, D2, D3 and D5 have acted as such zones of stress concentration due to increase in pore pressure by diffusion of water from the reservoir. The fact that the seismic activity from 1983 to 1990 has not spread beyond the block defined by these dykes points towards the possibility of these dykes acting as "barriers" and thus becoming a zone of high pore pressure.

If the seismic activity is noticed immediately after the impounding, then it can be interpreted in terms of increased pore pressure along the local zone of stress concentrations due to compression. If the seismicity occurs after a delay of several years, as in the case of Bhatsa, it can be attributed to diffusion-controlled changes in pore pressure causing failure along stressed zones.

The rate of loading also plays a major role in triggering the seismic activity. For example in the month of June, in 1983 the reservoir level of Bhatsa was raised by 18 meters. Such a steep rise in rate of filling may lead to compression-induced increases in pore pressure and subsequently to fault weakening. There will be an increase in pore pressure if the loading is rapid in comparison to the time constant for diffusion out of the zones of high initial pore pressure. On the contrary if the loading rate is gradual the pore pressure within the localized zone will dissipate before reaching critical values. Thus, the initial trigger in the Bhatsa earthquakes seems to have been caused by the rapid rate of loading in June, 1983.

Conclusions

The seismic activity near the Bhatsa reservoir was initially triggered along a NNE trending dyke (D2). The seismicity during 1983 to 1990 was confined to a block limited by dykes D1, D2, D3 and D5, which are likely to act as "Barriers" for the diffusion of water and may thereby become zones of stress concentration due to the increase in pore pressure causing failure along them.

Although the predominant mechanism of the Bhatsa earthquake sequence was pore pressure increase due to the diffusion of water to the hypocentral depth. The initial triggering of seismicity was due to an increase in pore pressure caused by elastic stresses due to rapid rate of loading of the reservoir in July, 1983.

Acknowledgements

The author is thankful to Maharashtra Government authorities for supporting these studies. Maharashtra Engineering Research Institute provided data for these studies. The author is grateful to Dr. Harsh K. Gupta, Director of the National Geophysical Research Institute, Hyderabad for his keen interest and guidance in this study.

REFERENCES

BAPAT, A., KULKARNI, R. C., and GUHA, S. K. (1983), *Catalogue of Earthquakes in India and Neighbourhood from Historical Period upto 1979*, Indian Soc. Earthq. Tech. Roorkee, 211 pp.

CHADHA, R. K. (1991), *Studies on Reservoir-induced Seismicity for Some Cases in Peninsular India*, Ph. D. Thesis, Indian School of Mines, Dhanbad, Bihar India, 201 pp.

GUPTA, H. K., and RASTOGI, B. K., *Dams and Earthquakes* (Elsevier, Amsterdam 1976) 299 pp.

KAILA, K. L., MURTHY, P. R. K., RAO, V. K., and KHARETCHKO, G. E. (1981a), *Crustal Structure from Deep Seismic Sounding along the Koyna II (Kelsi-Loni) Profile in the Deccan Trap Area, India*, Tectonophys. 73, 365–384.

KAILA, K. L., REDDY, P. R., DIXIT, M. M., and LAZARENKO, M. A. (1981b), *Deep Crustal Structure at Koyna, Maharashtra Indicated by Deep Seismic Sounding*, J. Geol. Soc. India 22, 1–16.

LEE, W. H. K., and LAHR, J. C. (1972), *A Computer Program for Determining the Hypocenter, Magnitude and First Motion Pattern of Local Earthquakes, Revision of HYPO71*, U.S. Geol. Surv., Open File Report, 100 pp.

OFFICERS of GEOLOGICAL SURVEY of INDIA (1984), *A Note on Geology and Structure of Bhatsa-Khardi Area of Maharashtra in the Regional Perspective*. 12 pp.

PATIL, D. N., BHOSALE, V. N., GUHA, S. K., and POWAR, K. B. (1984), *The Khardi Earthquakes*, Current Sci. 53, 805–806.

RASTOGI, B. K., CHADHA, R. K., and RAJU, I. P. (1986), *Seismicity near Bhatsa Reservoir, Maharashtra, India*, Phys. Earth. Planet. Inter. 44, 179–199.

SIMPSON, D. W., LEITH, W. S., and SCHOLZ, C. H. (1988), *Two Types of Induced Seismicity*, Bull. Seismol. Soc. Am. 78, 2025–2040.

(Received May 30, 1994, revised/accepted December 20, 1994)

PAGEOPH, Vol. 145, No. 1 (1995)

0033–4553/95/010167–08$1.50 + 0.20/0

Speculation on the Causes of Continuing Seismicity Near Koyna Reservoir, India

PRADEEP TALWANI[1]

Abstract — The cause for continuous induced seismicity at Koyna is not well understood. A heuristic model based on various physical parameters observed at Koyna is being proposed to explain the ongoing seismicity. This model contains two essential elements: (i) Intersecting faults near Koyna provide means of stress build-up in response to plate tectonic forces. (ii) The annual reservoir loading cycle and changes in the ground water table perturb this stress build-up by an influx of pore pressure in a fluid infiltrated medium. Hence, the spatial and temporal pattern of the pore pressure distribution and the seismicity will be governed by the location and hydromechanical properties of the faults and fractures. The predictions of the model can be tested by comparing the temporal and spatial pattern of seismicity with the changes in lake level and water table.

Key words: Intersection model, continuing seismicity.

Introduction

Impoundment of reservoirs can induce seismic activity, which sometimes can be destructive. Although over 70 cases of reservoir-induced seismicity (RIS) are known, only eleven reservoirs have been associated with magnitude $(M) \geq 5.0$ earthquakes and only four of them with $M \geq 6.0$ (GUPTA, 1992). Of these, Koyna was associated with the largest earthquakes (M 6.3) and is also the site of ongoing seismicity, over 30 years after impoundment.

Impounding of the reservoir behind the Koyna Dam started in 1962 and the seismic activity which followed in 1963 continues. At most reservoirs, seismicity follows the initial impoundment or increase in lake levels, but after a period of time the seismicity tapers off. Koyna is unique in that earthquakes are continuing thirty years after the filling and large events continue to be felt. In 1993, more than 4000 events with $M \geq 1.7$ were recorded near Koyna (MERI, 1994). This includes two earthquakes with $M \geq 5.0$. The latest of the series of moderate events was a M 5.4

---

[1] Department of Geological Sciences, University of South Carolina, Columbia, South Carolina 29208, U.S.A.

earthquake on February 1, 1994. It caused a fair amount of damage, ground breakage and increased seepage in the Koyna Dam. Therefore, the question arises, why do earthquakes (many with $M \geq 4.0$) continue to occur at Koyna over thirty years after the observation of the initial seismicity. This feature is unique to Koyna. A speculative model is proposed, based on the model for intraplate earthquakes and our understanding of the mechanisms of reservoir-induced seismicity.

Intersection Model for Intraplate Seismicity

According to this model, proposed to explain intraplate earthquakes (TALWANI, 1988, 1989), the earthquakes lie at the intersections of pre-existing zones of weakness. One of these zones of weakness is regional in scale, the intersecting zone may be local in nature. The intersections form the locked area of a fault and are locations of stress build-up and subsequent earthquakes (TALWANI, 1988). The intersections also help to localize seismicity (KING and NEBELEK, 1985). In a numerical model of two-dimensional static plane strain, ANDREWS (1989) noted that a bend in a fault (or an intersection) acts as a barrier and leads to stress concentration. This stress concentration occurs at the bend and will tend to induce slip on the spur. ELLIS (1991) demonstrated the validity of the model with *in situ* data from south central Oklahoma and MA *et al.* (1990) its validity by photoelastic experiments. From the seismicity data (TALWANI and RAJENDRAN, 1991) and theoretical considerations (ANDREWS, 1989) it is seen that the main shock may occur away from the intersection, whereas low-level seismicity is observed in the vicinity of the intersection itself.

Koyna Dam and Reservoir

Koyna Dam was built in 1962 across the 60-km N–S stretch of the Koyna river which flows in a 500–600-meter deep valley. South of Koyna Dam the river turns east near Helwak (Fig. 1). The dam is built on the basalt flows that form the Deccan traps. Several papers have been written which describe the seismicity and other aspects of induced seismicity. GUPTA (1992) presents a comprehensive review.

Two observations which are pertinent to the proposed model are the presence of two trends of faulting and the annual pattern of seismic activity. Evidence for two deformational trends comes from seismicity, geomorphic, lineament and meizoseismal data for the main shock. Fault plane solutions for the main shock (M 6.3 on December 10, 1967) were obtained by several workers (see SINGH *et al.*, 1975 for review; LANGSTON, 1976; GUPTA, 1992). They suggest that it was associated with the left-lateral strike-slip faulting on a generally steep plane striking between N10°E–S10°W and N26°E–S26°W. Focal mechanism for a large aftershock (M 5.3, December 12, 1967; RAO *et al.*, 1975) and foreshock (M 4.0–4.5, Septem-

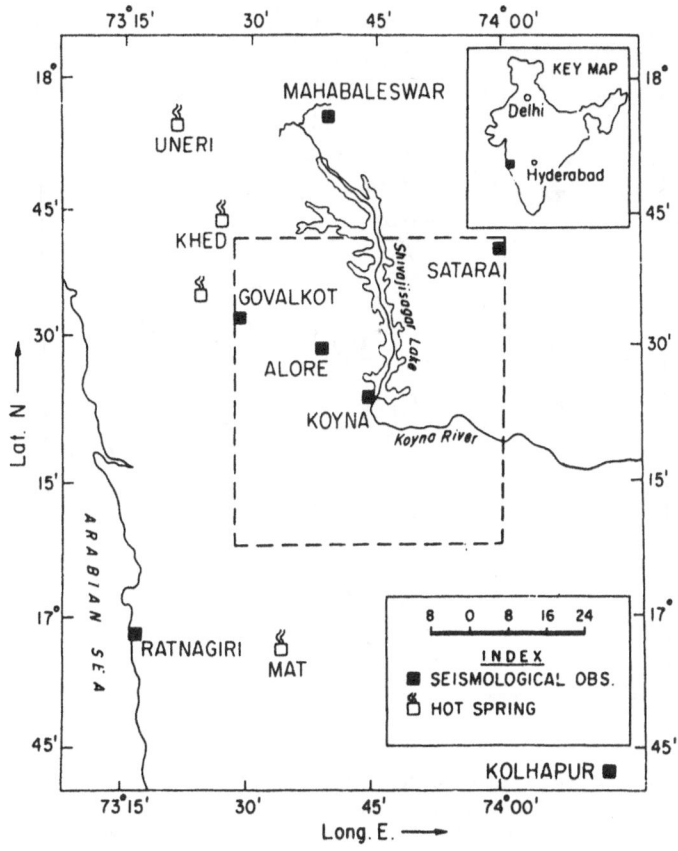

Figure 1

Location of Shivajisagar Lake formed by Koyna dam and the location of permanent seismic stations. Note the change in the direction of Koyna river, south of Koyna dam (from GUPTA et al., 1980).

ber 13, 1967; LANGSTON, 1981) are also similar to the main shock. For the M 5.3 aftershock of December 12, 1967, LANGSTON and FRANCO-SPERA (1985) obtained normal faulting on a plane striking N80°W ± 20°; dip 40° ± 10°. The exact location of this event is uncertain. Normal faulting on NW–SE striking planes was also obtained in composite fault plane solutions (CFPS) for earthquakes between 1967 and 1973 by RASTOGI and TALWANI (1980) and for some $M \geq 4.0$ events obtained from CFPS which ranged between N2°W to N67°W.

These data suggest two general trends of faulting, NNE–SSW and NW–SE, that are also seen on LANDSAT data and various geomorphic features. The origin of the intersecting faults is not clear. It is probably related to local and regional tectonothermal processes. Of interest here is their behavior in the present-day stress environment. Ground breakage that was observed following the main shock also trended NNE–SSW whereas the isoseismals for the largest damage in the meizoseismal area trended NW–SE (JOSHI, 1971). The isoseismals outside the meizoseis-

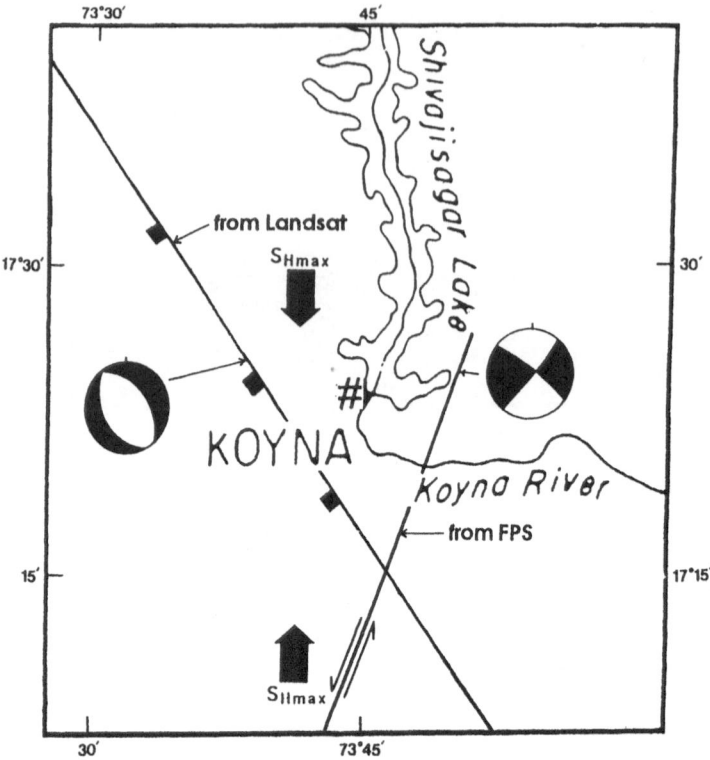

Figure 2

Schematic model of faults near Koyna. The main shock occurred on the NNE–SSW trending fault by
left-lateral strike-slip motion. Secondary faulting occurs on the NW–SE trending fault. The square tooth
pattern indicates normal faulting. Solid arrows indicate the direction of maximum horizontal compression.

mal area trended NNE–SSW (GUPTA, 1992). According to JOSHI, "The damage to
well-built structures in the village was observed and taken as a measure rather than
damage to poor type of construction." These observations and the abrupt change
in course of the river from N–S to E–W suggested the presence of two or more
faults (see e.g., RASTOGI and TALWANI, 1980; GUPTA *et al.*, 1980; LANGSTON and
FRANCO-SPERA, 1985).

In summary, based on the data presented above, there are faults along two
intersecting trends, $\sim$ N15°E–S15°W and the other roughly NW–SE. Seismicity on
the N15°E–S15°W is primarily by left-lateral strike-slip motion, whereas on the
NW trending fault it is by normal faulting. These styles of faulting are consistent
with the intersecting model for intraplate earthquakes, with the plate tectonic
stresses oriented roughly N–S (Fig. 2).

The seismicity at Koyna has shown a remarkable relation to its filling history.
The lake is filled between June and August annually during the rainy season and for
the rest of the year the water level decreases. The seismicity seems to follow the

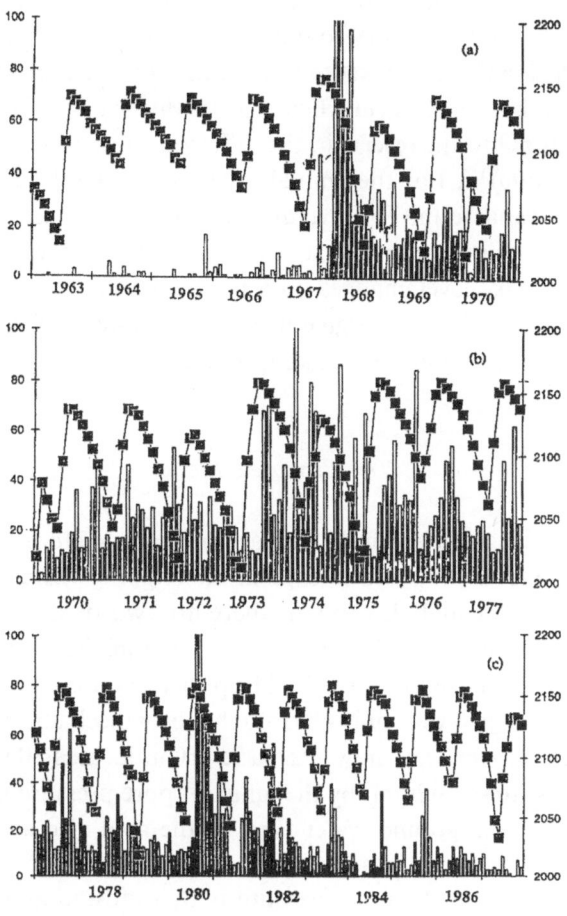

Figure 3
Seismicity (bars; $M > 2.0$) and water levels in feet (squares) at Koyna (a) from 1963 to 1970; (b) from 1970 to 1977 and (c) from 1978 to 1986. Number of earthquakes (left) and water levels in feet (right).

period of lake filling and the larger events occur 6–8 weeks after filling starts. This pattern has continued (Fig. 3).

Implications of Roeloffs' Model

In a model proposed to explain the seismicity associated with the cyclic fluctuation of a lake, ROELOFFS (1988) showed that the seismicity depended, among other factors, on the reservoir dimensions, the hydrological properties and the frequency of the lake level fluctuations.

ROELOFFS (1988) calculated the stress and pore pressure changes produced by cyclic loading in a reservoir. She found that the stress and pore pressure fields produced by cyclic loading of a reservoir of width L are governed by a dimensionless frequency Ω, given by $\Omega = L^2/2c\omega$ where c is the hydraulic diffusivity and ω is the frequency. Directly below the reservoir, the diffusion effect is negligible below a depth z given by $z = \pi(2c/\omega)^{1/2}$. For the annual cycle of lake level fluctuations at Koyna and $c = 1 \text{ m}^2/\text{s}$, (a value found for several reservoirs associated with RIS, TALWANI and ACREE (1984)), this corresponds to a depth of about 25 km. Thus, the effect of the annual cycle of lake level changes on a fault zone can reach several kilometers. It is anticipated that modeling will give us a better estimate of the effect of loading, away from the reservoir along pre-existing fractures.

The Proposed Model

Figure 2 shows the geometry of the faults and associated focal mechanisms. Based on the above, a model is proposed to explain the ongoing seismicity at Koyna. This model contains two essential elements, i) there are two or more NW–SE trending faults intersecting the major NNE–SSW to N–S feature. These intersections (south of Koyna) provide a means of stress build-up in response to plate tectonic forces (inferred to be N–S to just west of N–S from the focal mechanism) and ii) the annual loading cycle of the Koyna (and now Warna reservoir located 30 km to SE of Koyna Dam) perturbs this stress build-up by the influx of pore pressure in a fluid infiltrated medium. The fluctuating ground water table in the area also acts in the same way as the reservoir, although the changes are smaller.

The largest increase in the pore pressure is by diffusion and it is delayed with respect to the loading cycle. The effect of this increased pore pressure is to enhance the seismicity. The delay of 6 to 8 weeks is consistent with inferred permeabilities of fractured rocks. Fluid flow is considered unimportant in triggering the seismicity as the subsurface is assumed to be saturated. However, fluid flow can lead to chemical changes through dissolution of barriers, changes in permeability due to deposition or removal of particles, changes in coefficient of friction due to wetting of clay, etc. The long-term chemical effects due to fluid flow are not considered here. After the magnitude 6.3 event in December, 1967, and before the M 5.4 February, 1994 event, earthquakes with magnitude 5.0 or greater have occurred in October 1968, October 1973, and September–October 1980 (GUPTA and IYER, 1984). Thus the time period to reload the stresses required to generate these moderated ($M \geq 5.0$) events has increased with each episode. This observation suggests that the region was highly stressed in 1967 and that after each moderate earthquake it takes longer to build up the stresses to the levels necessary for $M \geq 5.0$ events. It is not clear whether the apparent 6.5 or 13 year cyclicity in the frequency (Fig. 3) of earthquakes of magnitude ≥ 5.0 is real and related to a discernible cause. We propose to test this model by mathematical modeling of stress build-up at intersecting faults and the changes in pore pressure due to water level changes in the reservoir and the ground water table, superposed on the ambient stress field.

Acknowledgements

The author is grateful to Dr. Kusala Rajendran for discussions and for Figure 3. The author also wishes to thank an anonymous reviewer for thoughtful comments.

REFERENCES

ANDREWS, D. J. (1989), *Mechanics of Fault Junctions*, J. Geophys. Res. *94*, 9389–9397.

ELLIS, W. L. (1991), *Stress Distribution in South Central Oklahoma and its relationship to crustal structure and contemporary seismicity*. In 32nd U.S. Symp. on *Rock Mechanics*, 73–81.

GUPTA, H. K., *Reservoir Induced Earthquakes* (Elsevier, Amsterdam 1992).

GUPTA, H. K., and IYER, H. M. (1984), *Are Reservoir-induced Earthquakes of Magnitude >5.0 at Koyna, India Preceded by a Couple of Earthquakes of Magnitude >4.0*, Bull. Seismol. Soc. Am. *74*, 863–873.

GUPTA, H. K., RAO, C. V. R., RASTOGI, B. K., and BHATIA, S. C, (1980), *An Investigation of Earthquakes in Koyna Region, Maharashtra for the Period October 1973 through December 1976*, Bull. Seismol. Soc. Am. *70*, 1833–1847.

JOSHI, R. N. (1971), *Study of Damage and Throws in Koyna Earthquake*, Indian J. Power and River Valley Development, (Special Number), 39–45.

KING, G., and NEBELEK, J. (1985), *Role of Faults in the Initiation and Termination of Earthquake Rupture*, Science *228*, 984–987.

LANGSTON, C. A. (1976), *A Body Wave Inversion of the Koyna India, Earthquake of December 10, 1967 and Some Implications for Body Wave Focal Mechanisms*, J. Geophys. Res. *81*, 2517–2529.

LANGSTON, C. A. (1981), *Source Inversion of Seismic Waveforms: The Koyna India, Earthquakes of 13 September 1967*, Bull. Seismol. Soc. Am. *71*, 1–24.

LANGSTON, C. A., and FRANCO-SPERA, M. (1985), *Modeling of the Koyna India, Aftershock of 12 December 1967*, Bull. Seismol. Soc. Am. *75*, 651–660.

MA, A., FU, A., ZHANG, Y., WANG, C., ZHANG, G., and LIU, D., *Earthquake prediction—Nine major earthquakes in China (1966–1976)*, (Beijing Seismological Press Beijing and Springer-Verlag 1990).

MERI, (1994), *Thirteenth Koyna Tremors Sub-committee Meeting, Pune, 5 January, 1994*, Seismic Status Report of Koyna Region, Maharashtra Engineering Research Institute, Nasik, 27 pp.

RAO, B. R., RAO, T. K. S. P., and RAO, V. S. (1975), *Focal Mechanism Study of an Aftershock in the Koyna Region of Maharashtra State, India*, Pure and Appl. Geophys. *113*, 483–488.

RASTOGI, B. K., and TALWANI, P. (1980), *Relocation of Koyna Earthquakes*, Bull. Seismol. Soc. Am. *70*, 1849–1868.

ROELOFFS, E. A. (1988), *Fault Stability Changes Induced Beneath a Reservoir with Cyclic Variations in Water Level*, J. Geophys. Res. *93* (B3), 2107–2124.

SINGH, D. D., RASTOGI, B. K., and GUPTA, H. K. (1975), *Surface-wave Radiation Pattern and Source Parameters of the Koyna Earthquake of December 10, 1967*, Bull. Seismol. Soc. Am. *65*, 711–731.

TALWANI, P., and ACREE, S. (1984), *Pore Pressure Diffusion and the Mechanism of Reservoir-induced Seismicity*, Pure and Appl. Geophys. *122*, 947–965.

TALWANI, P. (1988), *The Intersection Model for Intraplate Earthquakes*, Seis. Res. Lett. *59*, 305–310.

TALWANI, P., *Characteristic features of intraplate earthquakes and the models proposed to explain them*. In *Earthquakes at North-Atlantic Passive Margins: Neotectonics and Post Glacial Rebound* (eds.

Gregersen, S., and Basham, P. W.) (NATO ASI Series. Series C. Mathematical and Physical Sciences *266*, 1989) pp. 563–579.

TALWANI, P., and RAJENDRAN, K. (1991), *Some Seismological and Geometric Features of Intraplate Earthquakes*, Tectonophys. *186*, 19–41.

(Received June 6, 1994, revised/accepted March 27, 1995)

PAGEOPH, Vol. 145, No. 1 (1995)

0033–4553/95/010175–18$1.50 + 0.20/0

Seismicity of the Koyna Region and Regional Tectonomagmatism of the Western Margin (India)

U. RAVAL[1]

Abstract — Recent findings on the Meso-Cenozoic tectonomagmatism and deep-seated anomalous geophysical structures suggest a close linkage between the seismicity of the Koyna region, the Western-ghat uplift (WG-U) and associated thermomechanical and fluid activities. The WG-U seems to be the result of late Cretaceous thermal mobilization, erosion of the Deccan trap cover and superposition of compressional stress. The association of seismicity with uplift seems to result from movement of deep-seated heat and fluids/volatiles along the edges (or boundary faults) of the uplift; because the force required for crustal deformation depends on the relief. Observed gradients in relief may be attributed to the differential erosion-rates and heat inputs, due to the time gap of ~ 50 Ma in the break-ups and plume activities on the eastern and western sides and consequence magmatism. Further, the geology and tectonics strongly indicate that the western margin (WM) is a relic of a mobile arm (MA), that included Madagascar, and which formed a part of the Proterozoic mobile belt of 'greater India' (for $t > 85$ Ma). The mobile nature of the WM facilitates mantle upwellings and transient elevation of isotherms at depth, raising the possibility of intermittent metamorphism and greater deformation.

Superposition of the ongoing compression and uplift-induced forces make local permeability and pore-fluid pressure vital in triggering the seismic slip over the Peninsular shield. Certain representative model calculations have been carried out to estimate change in the e.m. induction characteristics caused by an intermittent hydraulic connectivity. The results show a drop in the resistivity which could be a useful monitoring index. The close connection of uplift and fluid activity as discussed here seems applicable for other active parts of the South Indian Shield (SIS) also.

Key words: Mobile belt, uplifts, fluids, intraplate seismicity.

Introduction

It is nearly 3 decades since the intriguing Koyna earthquake, which occurred at the edge of the hitherto considered stable Indian shield. The analyses/investigations performed to date mostly deal with the parameterization of earthquakes (GUPTA *et al.*, 1969, 1973; JAI KRISHNA *et al.*, 1969; LEE and RALEIGH, 1969; SINGH *et al.*, 1975; LANGSTON, 1976; RASTOGI and TALWANI, 1980; PATIL *et al.*, 1986; GUPTA, 1992) and their temporal correlation with loading of the lakes/reservoirs (SINGH *et al.*, 1976; STUART-ALEXANDER and MARK, 1976). But relatively very few studies

---

[1] National Geophysical Research Institute, Hyderabad-7, India.

concern themselves with the tectonothermal nature of this region. The present work therefore endeavors to combine geodynamic history, geotectonic information and associated geophysical signatures over this area, which have a bearing on its seismicity. The seismicity associated with the reservoir impounding has been attributed to pore pressure changes resulting from (i) elastic compression, (ii) diffusion and (iii) response of the underlying elastic medium to loading of the reservoir. This implies that besides impounding of a reservoir, the geology and tectonics also play an important role. Hence the study of geodynamics, development of higher density of fractures (or conduits) and possible presence of fluids at depth are quite important to our understanding of the reservoir-induced seismicity (RIS). Hence the tectonomagmatic evolution since the Cretaceous time and its consequences such as uplifts, metamorphism, etc., seem to be among the primary causes of the seismic instability. An understanding of the development of these structural features and associated deep geophysical anomalies would vitally complement the parameterization/monitoring of seismicity not only in the Koyna area but also of the South Indian shield (SIS).

The Western Margin — A Mobile Arm

High uplift and/or subsidence, relatively large deformation, magmatism, heat flow and hotsprings along the volcanic western margin (WM) from Palghat to Kutch strongly suggest that the corridor consisting of the WM is mobile relative to the stable and deep-rooted Dharwar Craton (DC). It also follows from the evolution of the WM since the Cretaceous. Thus, as has been proposed (RAVAL and VEERASWAMY, 1991) prior to 80 Ma the WM and Madagascar formed one of the major arms of the Proterozoic mobile belt of "Greater India." This mobile arm (MA), termed Madagascar Western margin mobile arm (MW-MA), surrounded the DC from the western side. It appears to link the southernmost charnokite MA (SGMA) with the junction of Cambay, Delhi and Satpura MAs near the Saurashtra Peninsula (Fig. 1). The Seychelles, which separated from India at ~64 Ma, could have been a part of the Satpura MA along the westward extension of Narmada lineament.

MAs, encircling the cratons, facilitate the continental break-ups (or rifting) despite the deep roots ($d > 200$ km) as inferred beneath the Dharwar craton of the Indian shield from tomography (IYER et al., 1989; ANDERSON et al., 1992). These episodically remobilized corridors (or MAs) represent pre-existing weak (or thin) zones through which mantle upwelling and/or perispherical spread (ANDERSON, 1994a) is easier. This makes the MAs vulnerable to magmatism. It is perhaps pertinent that near Koyna (at Satara) some geological evidence is reported to suggest recent volcanism (KARMARKER et al., 1993). On the basis of deep seismic/ electric soundings and gravity studies, significant crustal thinning has been inferred

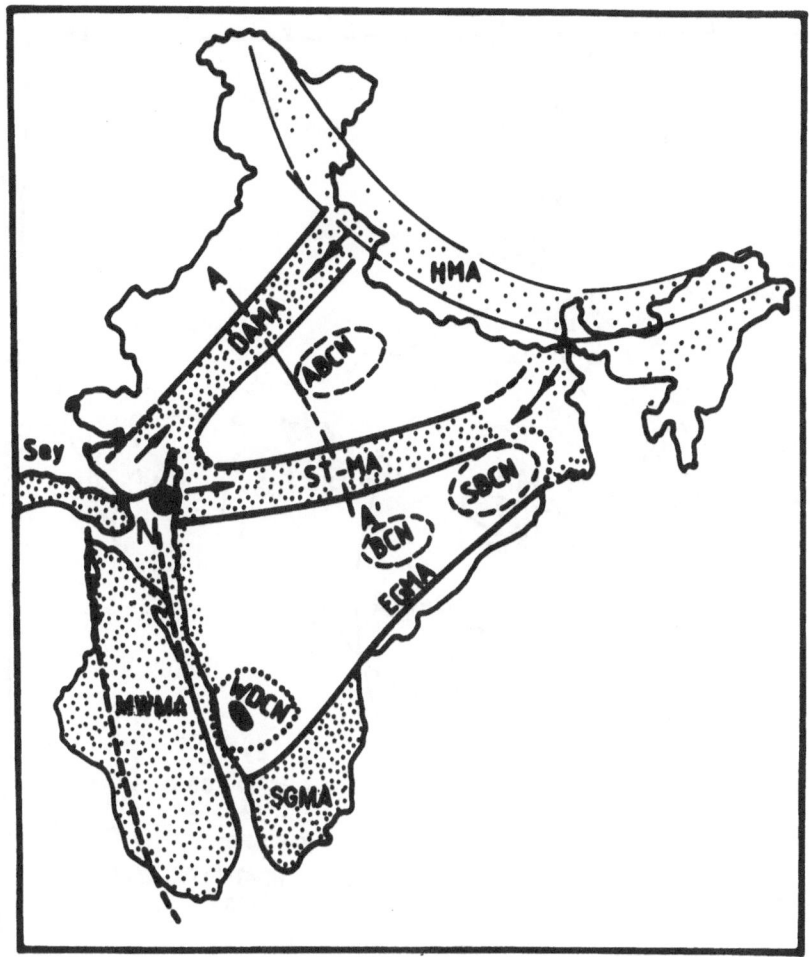

Figure 1

The paleogeography at $t = \sim 85$ Ma. The Madagascar and WM are shown as part of a mobile arm (MWMA) which surrounds the Dharwar craton from the western side. The other MAs are Delhi-Aravalli (DA-MA); Satpura (ST-MA), the Seychelles (Sey) may be part of this MA; and Easternghat (EG-MA). WDCN Western Dharwar Cratonic nucleii, SGMA—Southern Granulite mobile arm; HMA—Himalaya mobile arm; and N is the node where the MAs meet (BURKE and DEWEY, 1973). Along with the Bombay N represents one of major focii of Deccan Trap magmatism.

along WM from Cambay graben to Bombay (KAILA *et al.*, 1981, 1990; NEGI *et al.*, 1992; GOKARN *et al.*, 1992; ARORA and REDDY, 1991), which is a relic of the asthenospheric upwelling near the K-T boundary. Because of this, the crustal section beneath WM would have undergone large-scale fracturing/faulting as is manifested in uplift-associated hotspring, or hydrothermal system and heat flow (Fig. 2a).

Following JOHNSTON (1992) deformation in a MA such as WM (10^{-7}–10^{-8} s^{-1}) could be an order of magnitude greater than that within the stable Dharwar

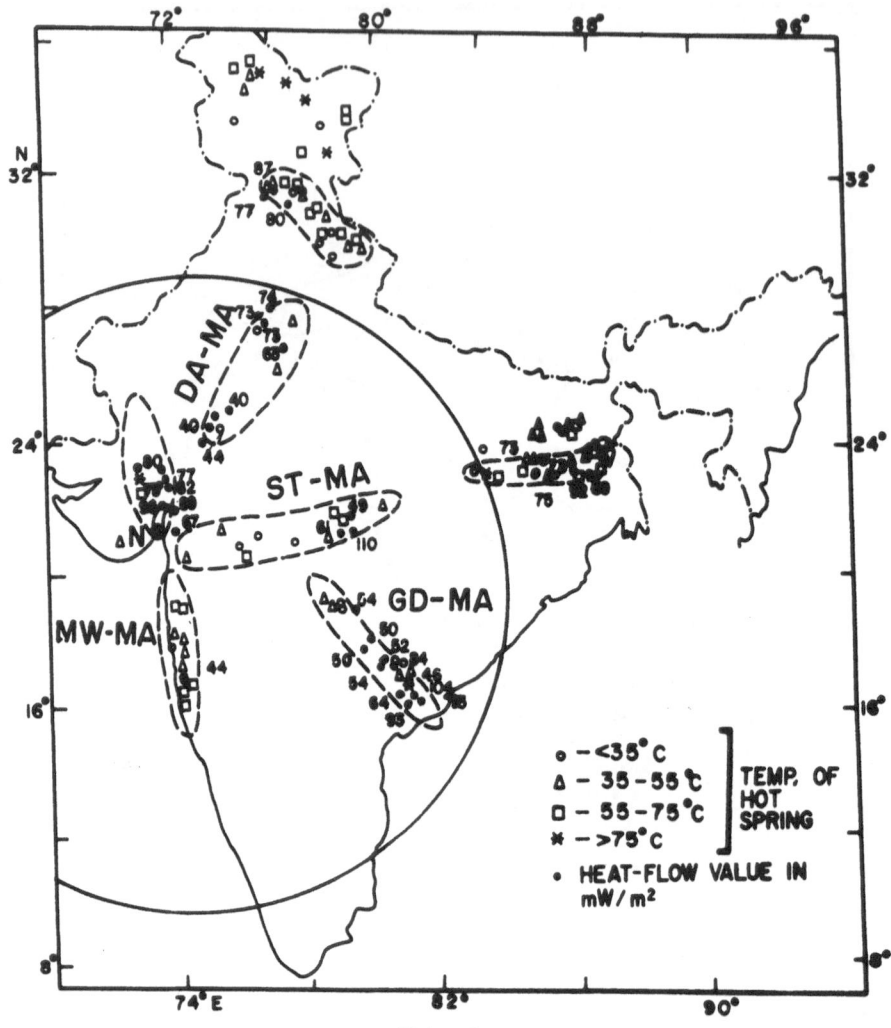

Figure 2a

Distribution of hotsprings and heat flow along the WM. The hotsprings seem to lie close to the edge of the uplifts and geological contacts. N is the node explained in Fig. 1.

craton (DC) which could be 10^{-8}–10^{-9} s^{-1}. Hence, higher seismic vulnerability (activity) is likely to manifest along the MAs (see Figs. 2b and 1). The deformations along MAs become further accentuated due to the regional compression resulting from (i) India-Asia collision and high plate velocity (~ 4.00 cm/yr) continuing over the past ~ 50 m.y., and (ii) simultaneous counterclockwise rotation. Since WM has been thermally mobilized thrice during the past 80 m.y. a far greater strain build-up occurs along it.

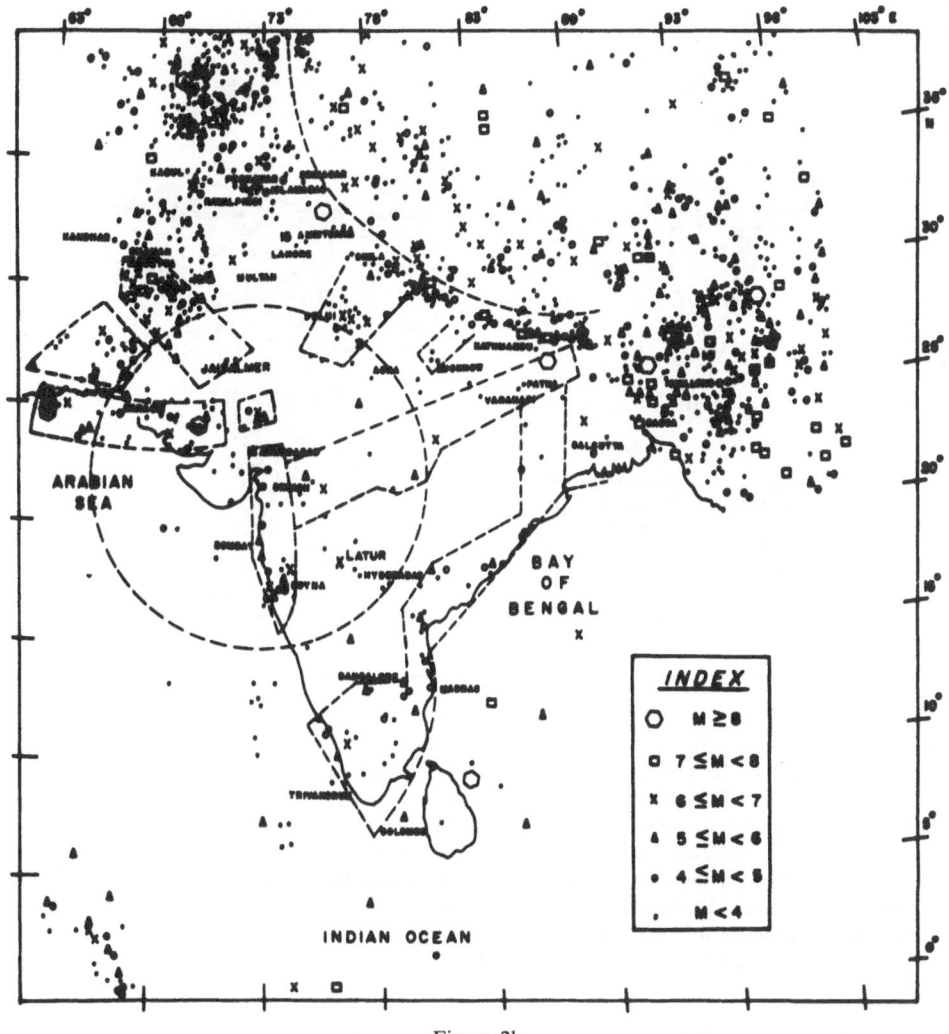

Figure 2b

Seismicity over the subcontinent (modified from KHATTRI *et al.*, 1984). A linear zone of seismicity extends from Kutch to Laccadive/Kerala along the western margin (WM). It shows a nearly radial pattern of seismicity spreading out from the Cambay node (*N*). Another pattern from NW to western margin runs almost linearly along the principal axis of an ellipse which includes Jaisalmer, Kutch, Cambay, Western Coast region (see Fig. 2). The higher level of seismicity along the MAs (shown in Fig. 1) indicates their different (or relatively vulnerable) rheology.

Uplifts and Morphotectonics

The SIS is a highland assumed to be uplifted recently (RADHAKRISHNA, 1993), may be grouped into (a) plateaus (Deccan and Karnataka) and the southernmost high grade terrain (see Fig. 3a) and (b) linear strips emerging mostly from the WM (Fig. 3b). This second category exhibits three patterns: The first one over the

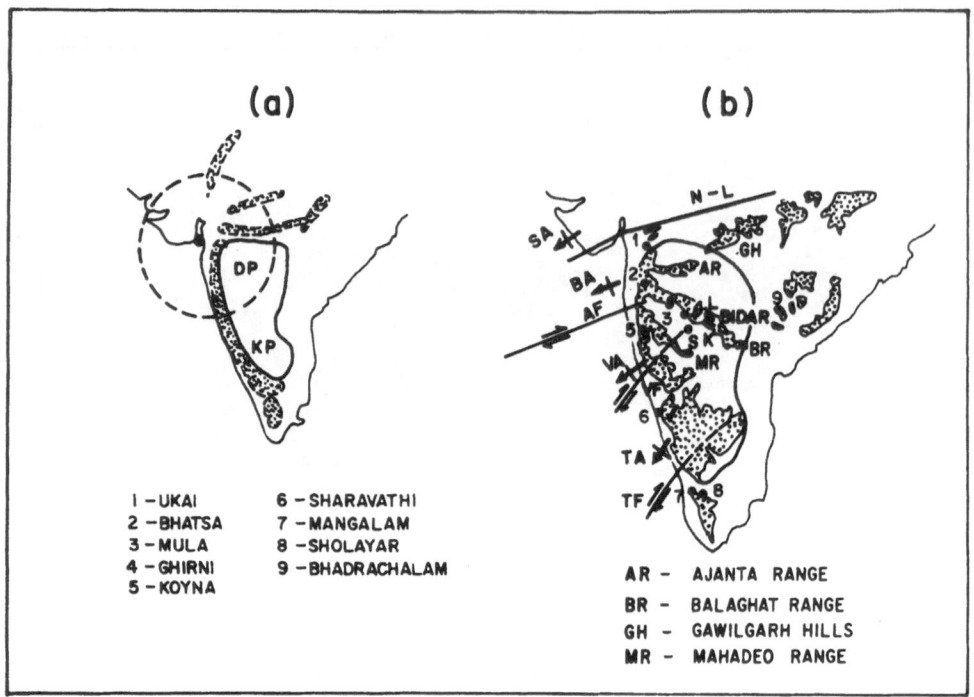

Figure 3
(a) The major radial spreads of the uplifts from the Cambay and Bombay nodes of successive plume outbursts, Westernghat, Delhi-Aravalli, Satpura horsts and Deccan (DP) and Karnataka (KP) plateau are depicted. (b) The uplifted blocks emerging from the Cambay-Bombay nodes of plume bursts. Over the Deccan plateau these are oriented mostly in NW–SE direction. The Mahadeo hills meet the WG-U near Koyna, while on the Balaghat Range (BR) the Killari (Latur) region lies. Structural features like archs/faults delineated on the seaward side, forming knots with the WG-U, are Narmada (N), Alibag (A), Venugurla (V) and Tellichery (T). DP—Deccan plateau; KP—Karnataka plateau; SA—Saurashtra arch; N-L—Narmada lineament; BA—Bombay arch; AF—Alibag fault; VA (or VF)—Venugurla arch (or fault); TA (or TF)—Tellichery arch (or fault).

Deccan plateau has, NW–SE orientations (e.g., Ajanta, Balaghat and Mahadev Ranges). The Mahadev hills intersect the WG-U close to Koyna; while within the Balaghat range (BR) lies the Killari (Latur) earthquake zone. The Karnataka plateau comprises the second category of uplifted blocks which are NE–SW oriented. The last is the N–S trending southernmost part. The three groups/zones are separated by major shear zones, lineaments and metamorphic transitions and also control the drainage system. Figure 3 displays these uplifted strips, their elevation varying from ∼600 to 2500 m. The Westernghat uplift (WG-U) along the WM is perhaps the most prominent due to its length and high relief. These uplifts (including basement ridges/highs) seem to be closely associated with the seismicity of the SIS, particularly the area covered by the DT and at the boundary (or edges) of uplifts. It follows from a nearly radial pattern of uplift spreading from the Cambay-Bombay node (Fig. 3a), that the thermal mobilization due to plume burst at ~ 67 Ma may be among the primary causes of the uplift. Other possibilities for the uplifts are:

(a) the WG-U could also be the result of shoulder uplift due to secondary convection (BUCK, 1986). The uplift of SGMA (Fig. 1) would also be affected similarly.

(b) The WG-U is also supported by the differential buoyancy. The latter results because the thermal remobilization of the WM occurred ~ 50 m.y. after that of the eastern margin. These thermal influxes during Cretaceous also may have led to the relatively hotter nature of the Indian shield (RAO and JESSOP, 1975; SINGH and NEGI, 1982; NEGI et al., 1986; RAVAL, 1993c).

(c) Erosion of the relatively high density basaltic cover of the DT, which initiates the buoyancy force controlled by viscosity at depth. The uplifts also mobilize crustal fluids, as evidenced from the hotsprings which occur at the edge of the area covered by the DT (see Fig. 1, CHADHA, 1992) where the buried faults may have been exposed after DT-erosion.

(d) Superposition of compression over the continental crust-lithosphere weakened at depth by heat and fluids from rifting and magmatism, especially within MAs. The resulting ductile lower crust exerts an upthrust over the overlying brittle layers.

(e) Magmatic underplating due to Reunion plume also gives rise to uplift (MCKENZIE, 1984).

The fractures/faults developed consequent to uplift and large-scale doming (COX, 1989) result in deep-reaching conduits which allow heat and mass transfer. The observed gravity field pattern (Fig. 2 in QURESHY, 1981; MISHRA, 1989; KAILASAM, 1993) corroborates the uplift. The long-period morphotectonics of regions between these uplifts therefore deserve a detailed analysis (MERRITTS and HESTERBERG, 1994).

Transitions in the Gravity and Elevation

A transition zone (TZ) exists along the WM over which major change in the elevation (E) and Bouguer gravity anomalies (Δg) is observed (Fig. 4). Thus north of TZ, the Δg is mostly high (~ -20 mgal) up to Jaisalmer while to its south Δg is quite low (in excess of -110 mgal). On the other hand, E is high (reaching >2000 m at many places) south of TZ, but becomes quite diminished to its north. The transitions of E and Δg would have structural implication. The gravity high north of TZ indicates magmatic underplating and/or crustal accretion (FYFE, 1993; STEIN and HOFMANN, 1994) in the mobile region, e.g., grabbroic intrusion which results in subsidence. Also gravity low to the south sugggests that, due to cratonic nature, perhaps only slight crustal accretion has occurred; instead, the presence of

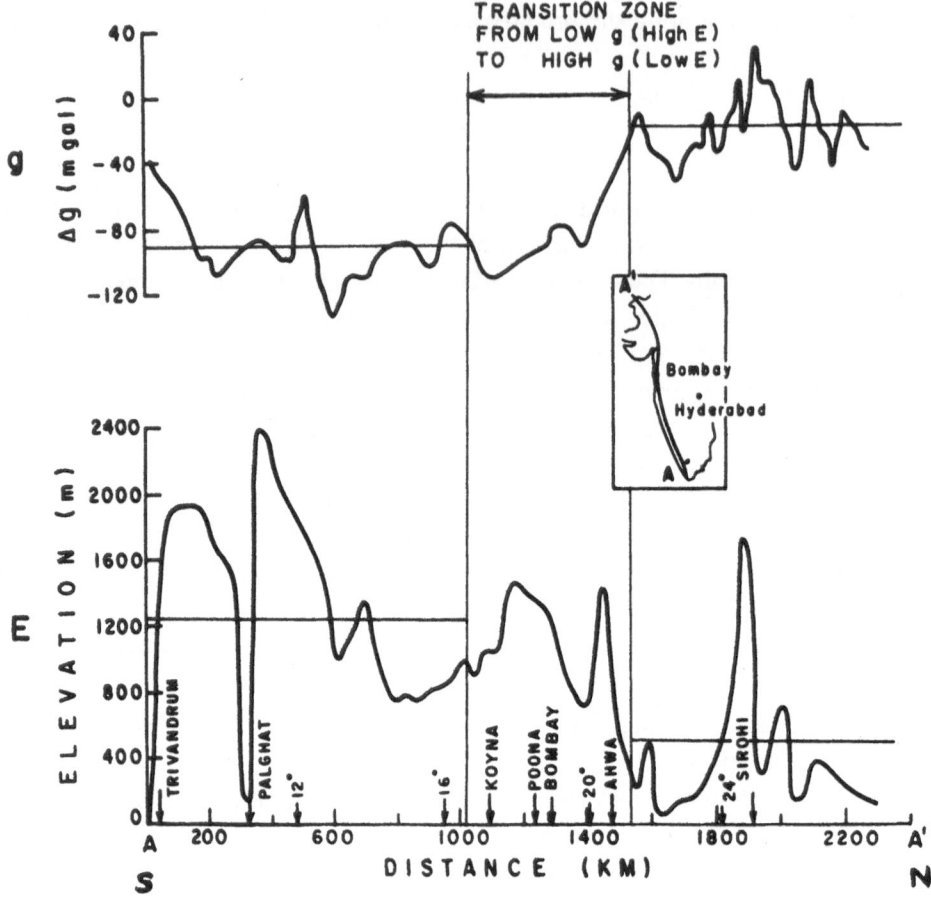

Figure 4

The Transition Zone (TZ) along the WM. It approximately comprises the Bombay to Koyna region. North of TZ the elevation (E) is low and Bouguer gravity anomalies (Δg) are high. But south of TZ the average value of 2 times g is low and that of E is high. Koyna lies near the southern end of TZ.

fluids released in magmatic/metamorphic process would have slightly reduced the density of the rock matrix.

'Transient' Metamorphism?

The crust beneath the WM would have received substantial thermal inputs and fluids due to (i) break-up from the Madagascar (~ 80 Ma), (ii) outburst of the Reunion plume (~ 67.5 Ma), associated alkaline and Deccan volcanism, (iii) break-up from the Seychelles at ~ 64 Ma and (iv) very closely placed trail of the plume (between ~ 65–58 Ma), and (v) magmatic underplating. Hence, underlying geotherms would be elevated for at least short periods and (P, T) conditions would be conducive to the prograde metamorphism, resulting in the release of fluid/ volatiles ($H_2O + CO_2$ system) in the deep crust (Fig. 5) and possibly a secondary replenishment of crust with mineral solutions. With regard to the activity of mineral solution, it is pertinent to point out that the depth of the seismogenic region (or brittle-ductile transition) lies close to the most common (P, T) range for gold concentration. Furthermore, a multi-stage and long-lived activation process is required for the development of faults/shear zones (or tectonic conduits) which are needed for the genesis of gold deposits.

Significantly, SIBSON (1987) has suggested that ruptures during earthquakes may aid the mineralization process. The deep water due to metamorphism/antexis would be expelled due to compression (MUIR-WOOD and KING, 1993). Extension-related metamorphism has been described by SANDIFORD and POWELL (1986). As described by BAILEY (1994), a part of these fluids could be trapped at intermediate crustal depths. Such release and trappings of fluids explains the low velocity (KRISHNA et al., 1991) and high conducting layers (GOKARN et al., 1992; SARMA et al., 1994) inferred in and around the Koyna region (see Fig. 6). The presence of fluids affects the spatial disposition of brittle-ductile transitions. Additionally, the thermally-induced ductility in the deep crust also exerts stress on the overlying brittle part. Thermal heating/cooling respectively following the uplift and erosion of the Westernghat would give rise to dehydration and hydration reactions similar to those described by COOMBS (1993) regarding low-grade mineral veins. The pore-fluid pressure (Pf) during these periods fluctuates, sometimes falling from hydro-static to lithostatic conditions and hence could act as a triggering agent.

In view of enhanced fluid and thermal activity due to advection and metamorphism, it is interesting to note that zeolitization exists as a band along the WM (JEFFERY et al., 1988). An upward movement of deeper fluid and/or mineral solutions would trigger hydrofracturing and establish an intermittent hydraulic connectivity between near-surface and deeper fluid systems (RAVAL, 1993a). Through deep and permeable faults the meteoric or surface derived water may percolate to 8–10 km depths (NESBITT and MUEHLENBACK, 1991; JONES et al., 1992; MORRISON, 1994). In view of this, it is significant that the RIS has been observed generally in the close vicinity of uplifts along the WM (Fig. 4) and gravity low.

Certain Local Factors in the Koyna Region

Along with its regional setting, additional local factors may be responsible for the 'extra' instability (CHADHA, 1989). From the thermomagmatic evolution as outlined above, the development of such structural characteristics follows.

Transition Zone (TZ): As seen from Figure 4 the Koyna region lies within a major transition of the elevation (E) and Bouguer gravity anomalies (Δg). North of

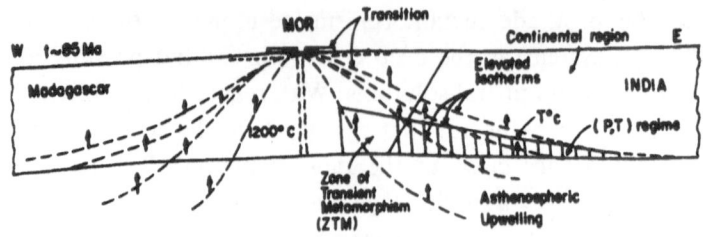

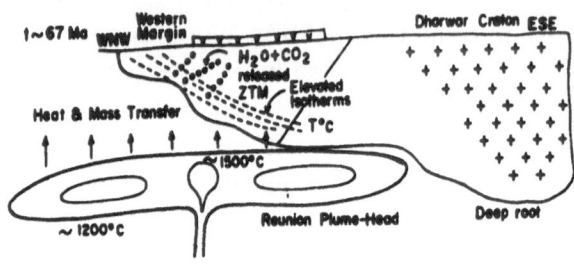

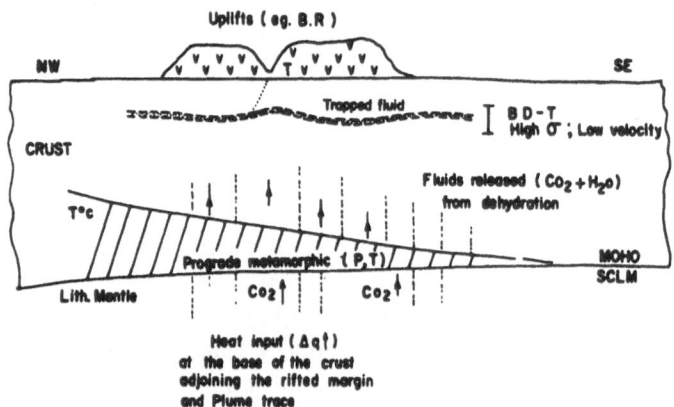

Figure 5

Modification (or elevation) of the geotherms, thermal influx and release/trapping of fluids in the deep crust and upper mantle. Top: near the break-up region between India and Madagascar (at ∼80 Ma), Middle: outburst of the Reunion plume head at ∼67 Ma, and Bottom: beneath an uplift such as Balaghat range (BR see Fig. 4b) which is closely associated with Killari (Latur), Hyderabad, etc. The fluid system (CO_2 and H_2O) from the metasometer magmatic and metamorphic processes may get trapped in crust forming high conductivity and low velocity layers and showing a certain degree of brittle-ductile transition (BD-T). Possible permeable conduits developed at the edge of an uplift for percolation of surface-derived (meteoric) water is shown by a thin line. MOR—Mid-oceanic ridge across which India and Madagascar split; BD-T—Brittle-ductile transition; SCLM—subcontinental lithospheric mantle.

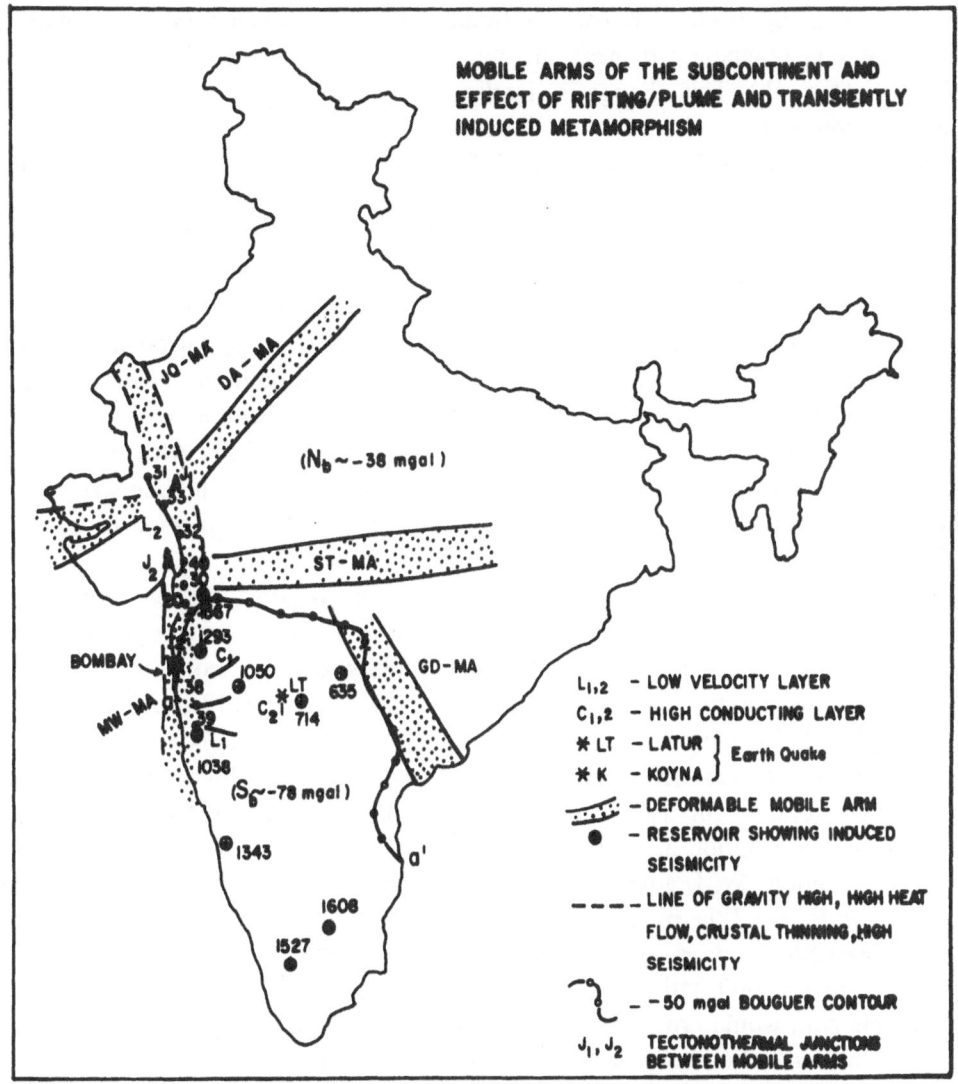

Figure 6

Location of low velocity layers (L_1, L_2) anomalously high conducting layers (C_1, C_2) over the Koyna and adjoining regions of DT-covered areas. Reservoir-induced seismicity cases are also shown; the number in brackets is value (in meter) of the nearest high relief. Line of thinning, high-heat flow, hydrothermal activity and hotsprings, seismicity and thermal gradients, etc. are shown along with the deformable MAs (dotted region) of the Indian subcontinent. The contour aa' broadly divides the subcontinent into low (− 78 mgal) and high (− 38 mgal) gravity zones.

TZ the average E is small (~ 400 m) and Δg high (~ − 20 mgal), while to its south E reaches a high average value (~ 1200 m) and Δg falls to low magnitudes (~ − 90 mgal). This implies additional mechanical strain.

Structural Joints: The Koyna region lying on the eastern slope of the WG-U is being approached from either side by almost transversely oriented structural features (Fig. 3b). South of Koyna the trend of NW–SE directed uplifts emanating from the WG-U changes to near east-west and slightly further south it becomes NE–SW (Fig. 3b). This change is significant for local stress build-up in view of ongoing compression. In addition, according to BISWAS (1993) offshore explorations have delineated certain ENE–WSW oriented arches and strike-slip faults (viz., Alibagh, Venugurla, and Tellichery), on the seaward side of WM extending into the continental interior. They would also form knots with the WG-U and other uplifted blocks (Fig. 3b). Regions of hydrothermal system and higher seismicity appear to lie near these knots (e.g., Ganeshpuri, Koyna, Latur, etc.).

At the resulting structural junctions, partitioning of stresses, crustal weakening, development of more fractures and conduit permeability occur. These factors become further significant in light of a recent suggestion of Beaumont as quoted by FYFE (1993), according to which heavy rainfall in the Southern Alps of western New Zealand and stress field play a controlling role in shaping the fine structures. The WM also receives a relatively large amount of rains in comparison to the rest of the platform.

Exit of Plume to Ocean: As seen in Figure 2a, maxima of heat input and hydrothermal activity are observed along the WM. Figure 7 depicts the possible surface manifestations (topography, tectonic, geophysical and geological) resulting in 80 to 60 Ma, due to passage of the WM over the plume head (SIVARAMAN and RAVAL, 1992). The elliptical envelope (thin line) of plume pulses comprises regions of high seismicity (or deformation), hydrothermal activity, high heat flow and thermal gradient, magmatism, transient metamorphism, uplift/subsidence, mobilization of mineral solution in crust, etc. Since the principal axis of the elliptical path lies on the mobile WM, additional strain would occur at the embedded structural joints. If in Figure 7 the picrite basalt at Rajpipla (RP) lies at the axis of the plume head (CAMPBELL and GRIFFITHS, 1990; ANDERSON, 1994b), and the Laccadive (L) are its oceanic manifestations at 60 Ma, then between these two space-time events the plume path would cross the WM from the continental to oceanic sides. On the basis of the correlation between the topographic relief (E) and Bouguer gravity high (Δg), along the WM with the plume trace, it follows that after the major hotspot outburst close to Bombay, the trail of the plume moved into the ocean through the southern end of TZ, i.e., near Koyna (Fig. 4).

Models of the Fluid Associated Changes in Electrical Resistivity

A number of factors indicate the possibility of fluids and fluid-filled conduits developing along the WM and also in the DT region, especially in areas of tectonic joints (e.g., south of Koyna). Currently, electrical conductivity is one of the most sensitive parameters to the presence of fluids and temperature (WENZEL and SANDMEIER, 1992; GOUGH, 1992; JONES, 1992; HYNDMANN et al., 1993) because

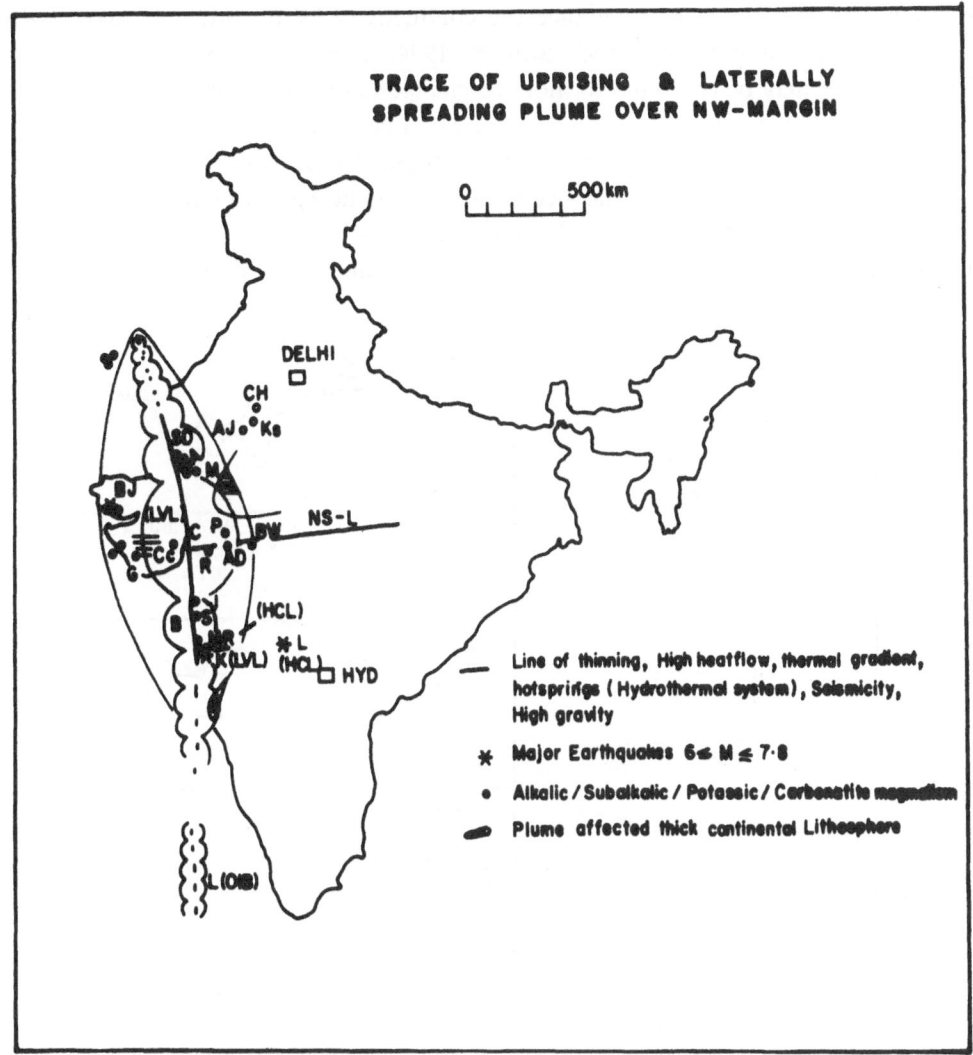

Figure 7

Possible plume trace caused by the NW part of the subcontinent due to interaction of the Reunion plume developing beneath the fast moving Indian lithosphere. It shows gradual spreading of the plume over the NW phase due to (i) uprising of the plume in the WM and (ii) northward motion and counterclockwise rotation of the Indian plate. The places marked (○) are locations of alkalic, subalkalic and potassic magmatism; those marks (*) represent large earthquake ($6.7 < M < 7.8$); The elliptical envelope of the plume path contains a region of relatively high seismicity, especially along its major axis (compare with Fig. 2b). This is also the line of crustal thinning, gravity high, heat flow, high thermal gradient, hydrothermal system and other associated geophysical, geochemical features observed here.

even slightly 'wet' crust can enhance the conductivity by an order of magnitude (SHANKLAND and ANDERS, 1983; BAILEY, 1994; YARDLEY and VALLEY, 1994). This property may be exploited to monitor change in the fluid volume and activity, due to alterations in reservoir level and increased permeability of the crust affected by compression (BRIGGS, 1993) and uplift-induced stresses. Major modification of the near-surface water regime has been observed at the Killari (Latur) earthquake region, which lies adjacent to the Balaghat uplift (BR in Fig. 3b) and exhibits a thrust (or reverse) faulting. In such situations the crust may behave like a

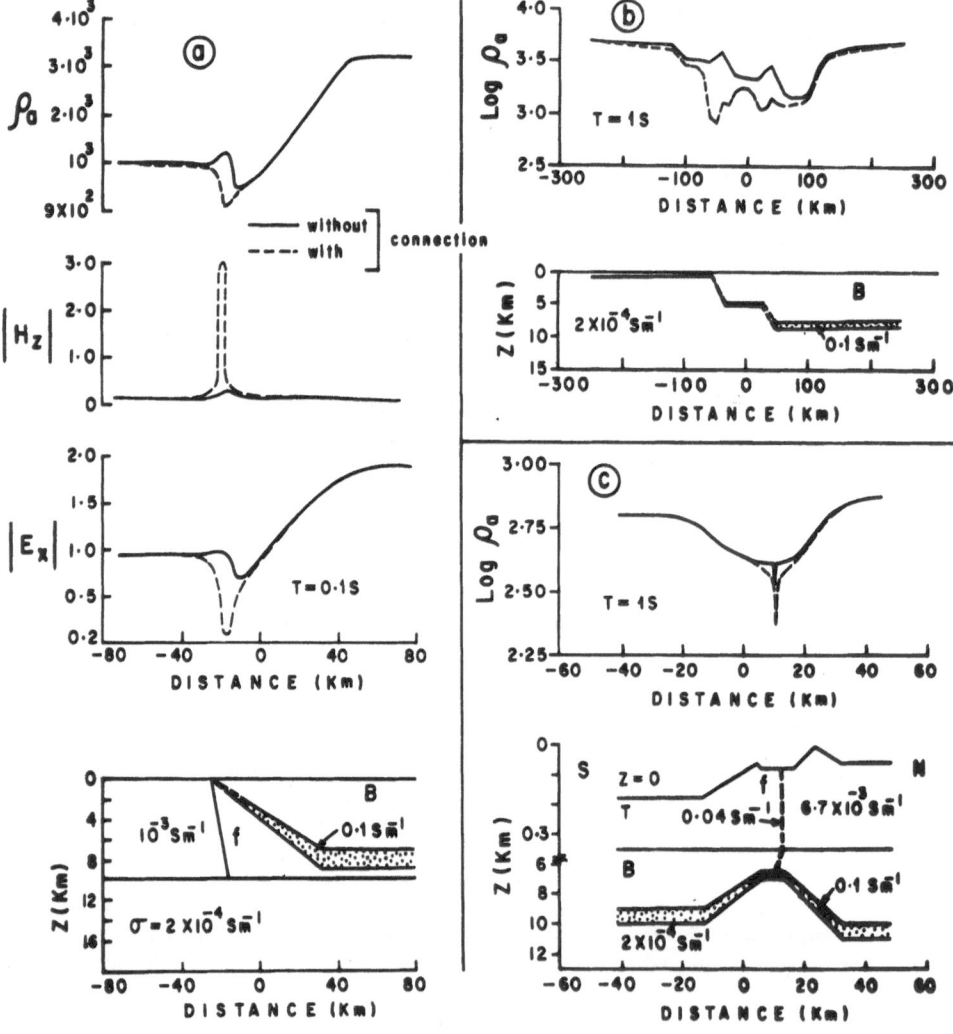

Figure 8

The changes in the induced electrical and magnetic field components (E, H) and apparent resistivity character with and without the connectivity between (a) the near surface and deeper layers, (b) more than one layer lying at varying depths in the crust, and (c) a model of Killari (Latur) region.

poroelastic medium (RICE and CLEARY, 1976). Thus it is important to estimate whether a difference in subterranean conductivity develops between the pre- and post-monsoon periods, and this is quantified by computing the electromagnetic induction response characteristics of certain representative models (Fig. 8). In this numerical analysis attention has been paid to the perturbation caused by the resistivity values due to the presence and absence of hydraulic connectivity between various depths. When an intermittent connectivity exists between fluid systems at different depths, the computed results clearly exhibit a 'drop' in the resistivity values.

The temporal variation of the conductivity structure thus assumes particular significance in estimating and understanding the role of young (shallow/external) and old (deep/internal) fluid regimes as the triggering agent. Both components may be present in regions like the Deccan Trap, where the crustal conductivity zone nearly overlaps with the focal region (SARMA et al., 1994). As noted earlier, this depth, interestingly, lies close to that for gold concentration due to favorable (P, T) conditions. Hence, chemical/isotopic analyses of mixed fluid regimes in conjuction with resistivity measurements and dynamic permeability could be a vital supplement to earthquake monitoring and associated processes.

Conclusions

The mobile nature of the WM has been described along with its tectonothermal development to explain the greater deformation. The uplifts caused by thermomagmatic and metamorphic inputs, erosion of DT, and superposition of compression over the MAs seem to provide a major source of seismic vulnerability due to advective heat transfer and gravity-driven fluid flow along the associated faults and depressions. Such correlation between seismicity and uplifts and fluid activity also seems to hold for other parts of the SIS. The study also outlines the role of seismically significant structural joints and tectonic transitions and other factors localized around Koyna; especially the enhanced permeability of conduits for deep percolation of surface-derived fluids. Therefore a detailed analysis of stream network connected with uplift would form an important input for monitoring the seismicity as has been done in the case of the New Madrid seismic zone (MERRITTS and HESTERBERG, 1994). Inferences of the deep geophysical probings and tectono-structural information support the possibility of a transient thermal metamorphism along the WM which could give rise to fluid at depth (water sills/aquitard, etc.). Model calculation shows that the monitoring of electrical resistivity changes due to fluid filling of interconnected rock matrices could be a useful surveillance tool.

Acknowledgements

The author expresses his thanks to the Director, NGRI for his kind permission to present the study at the IASPEI-94. He is grateful to Dr. R. K. Chadha for critically reviewing the manuscript and offering valuable suggestions. Sincere thanks are also extended to Sri K. Veeraswamy for multifaceted assistance. Mrs. Lakshmi Janakiraman and Sri K. Gopalakrishnan have also assisted in the preparation of the paper.

REFERENCES

ANDERSON, D. L. (1994a), *Lithosphere and Flood Basalts*, Nature Sci. Corres. *367*, 226.

ANDERSON, D. L. (1994b), *Komatiites and Picrites: Evidence that the 'Plume' Source is Depleted*, Earth and Planet. Sci. Lett. *128*, 303–311.

ANDERSON, D. L., ZHANG, T., and TANIMOTO, T. (1992). *Plume heads, continental lithosphere, flood basalts and tomography*. In *Magmatism and the Causes of Continental Break-up* (eds. Storey, B. C., Alabaster, T., and Pankhurst, R. J.), Geol. Soc. Spl. Publ. *68*, 99–124.

ARORA, B. R., and REDDY, C. D. (1991), *Magnetovariational Study over a Seismically Active Area in the Deccan Trap Province of Western India*, Phys. Earth Planet. Int. *66*, 118–131.

BAILEY, M. C. (1994), *Fluid trapping in Mid-crustal Reservoirs by H_2O–CO_2 Mixtures*, Nature *371*, 238–240.

BISWAS, S. K., *Tectonic framework and evolution of graben basins of India in Rifted Basins and Aulacogens* (ed. Casshyap, S. M.) (Gyanodaya Publ. Nainital, India 1993) pp. 18–32.

BRIGGS, R. O. (1993), *Effect of Loma Prieta Earthquake on Surface Water in Waddell Valley*, Water Resour. Bull. *27*, 991–999.

BUCK, W. R. (1986), *Small-scale Convection Induced by Passive Rifting: The Cause for Uplift of Rift Shoulders*, Earth and Planet. Sci. Lett. *77*, 362–372.

BURKE, K., and DEWEY, J. F. (1973), *Plume-generated Triple Junctions: Key Indicators in Applying Plate Tectonics to Old Rocks*, J. Geology *81*, 406–433.

CAMPBELL, I. H., and GRIFFITHS, R. W. (1990), *Implications of Mantle Plume Structure for the Evolution of Flood Basalts*, Earth and Planet. Sci. Lett. *99*, 79–93.

CHADHA, R. K. (1989), *Studies on Reservoir-induced Seismicity for Some Cases in Peninsular India*, Ph.D. Thesis, Indian School of Mines. Dhanbad, Bihar, India. 201 pp.

CHADHA, R. K. (1992), *Geological Contacts, Thermal Springs and Earthquakes in Peninsular India*, Tectonophys. *213*, 367–374.

COOMBS, D. S. (1993), *Dehydration Veins in Diagenetic and Very Low-grade Metamorphic Rocks: Features of the Crustal Seismogenic Zone and their Significance to Mineral Facies*, J. Metamorphic Geol. *11*, 389–399.

COX, K. G. (1989), *The Role of Mantle Plume in the Development of Continental Drainage Patterns*, Nature *342*, 873–877.

FYFE, W. S. (1993), *Hot Spots, Magma Underplating, and Modification of Continental Crust*, Can. J. Earth Sci. 30, 908–912.

GOKARN, G., RAO, C. K., SINGH, B. P., and NAYAK, P. N. (1992), *MT Studies across the Kurudwadi Gravity Features*, Phys. Earth. Planet. Int. *72*, 58–67.

GOUGH, D. I. (1992), *Electromagnetic Exploration for Fluid in the Earth's Crust*, Earth Science Rev. *32*, 3–18.

GUPTA, H. K. (1992), *Reservoir-induced Earthquakes*, Dev. in Geotech. Eng. *64*, Elsevier. 364.

GUPTA, H. K., NARAIN, H., RASTOGI, B. K., and MOHAN, I. (1969), *A Study of the Koyna Earthquake of December 10, 1967*, Bull. Seismol. Soc. Am. *59*, 1149–1162.

GUPTA, H. K., RASTOGI, B. K., and NARAIN, H. (1973), *A study of earthquakes in the Koyna region and common features of the reservoir-associated seismicity*. In *Man-made Lakes: Their Problems and Environmental Effects* (eds. Ackermann, W. C., White, G. F., and Worthington, E. B.), Am. Geophys. Union, Geophys. Monogr. *17*, 455–467.

HYNDMAN, D., VANYAN, L. L., MARQUIS, G., and LAW, L. K. (1993), *The Origin of Electrically Conductive Lower Continental Crust: Saline Water or Graphite?* Phys. Earth. Planet. Int. *81*, 325–344.

IYER, H. M., GAUR, V.K., RAI, S. S., RAMESH, D. S., RAO, C. V. R., SRINAGESH, D., and
SURYAPRAKASHAM, K. (1989), *High Velocity Anomaly beneath the Deccan Volcanic Province:
Evidence from Seismic Tomography*, Proc. Ind. Acad. Sci. (Earth and Planet. Sci. Section) *98*, 31–60.

JAI KRISHNA, CHANDRASEKHARAN, A. R., and SAINI, S. S. (1969), *Analysis of Koyna Accelerograms
of December, 11, 1967*, Bull. Seismol. Soc. Am. *59*, 1719–1731.

JEFFERY, K. L., HENDERSON, P., SUBBARAO, K. V., and WALSH, J. N. (1988), *The Zeolites of the
Deccan Basalt — A Study of their Distribution*. In *Deccan Flood Basalts* (ed. Subbarao, K. V.), J. Geol.
Soc. Ind. Mem. *10*, 151–162.

JOHNSTON, A. C. (1992), *Intraplate not Always Stable*, Nature *355*, 213–214.

JONES, A. G. (1992). *Electrical properties of the lower continental crust; in continental lower crust*. In
Developments in Geotectonics 23 (eds. Fountain, D. M., Arculus, R., and Kay, R. W.) (Elsevier 1992),
pp. 81–143.

JONES, A. G., KURTZ, R. D., BOERNER, D. E., CRAVEN, J. A., MCNEICE, G. W., GOUGH, D. I.,
DELAURIER, J. M., and ELLIS, R. G. (1992), *Electromagnetic Constraints on Strike-slip Fault
Geometry — The Frazer River Fault System*, Geology *20*, 561–564.

KAILA, K. L., KRISHNA, V. G., and MALL, D. M., (1981), *Crustal Structure along Mehmadabad-Bil-
limora Profile in the Cambay Basin, India from Deep Seismic Soundings*, Tectonophys. *76*, 99–130.

KAILA, K. L., TEWARI, H. C., KRISHNA, V. G., DIXIT, M. M., SARKAR, D., and REDDY, M. S. (1990),
Deep Seismic Sounding Studies in the North Cambay and Sanchor Basins, India, Geophys. J. Int. *103*,
621–637.

KAILASAM, L. N. (1993), *Geophysical and Geodynamical Aspects of the Maharashtra Earthquake of
September 30, 1993*, Current Sci. *65*, 736–739.

KAILASAM, L. N., MURTHY, B. G. K., and CHAYANULU, A. Y. S. R. (1972), *Regional Gravity Studies
of the Deccan Trap Areas of Peninsular India*, Current Sci. *41* (11), 403–407.

KARMARKER, B. M., KULKARNI, S. R., and MARATHE, S. (1993), *Quaternary Volcanic Activity in
Deccan Plateau*, Current Sci. *64*, 923–925.

KHATTRI, K. N., ROGER, A. M., PERKIN, D. M., and ALGERMISSEN, S. T. (1984), *A Seismic Hazard
Map of India and Adjacent Areas*, Tectonophys. *108*, 93–134.

KRISHNA, V. G., KAILA, K. L., and REDDY, P. R. (1991), *Low Velocity Layers in the Subcrustal
Lithosphere beneath the Deccan Traps Region of Western India*, Phys. Earth Planet. Int. *67*, 288–302.

LANGSTON, C. A. (1976), *A Body Wave Inversion of the Koyna, India, Earthquake of December 10, 1967,
and Some Implications for Body Wave Focal Mechanisms*, J. Geophys. Res. *81* (14), 2517–2529.

LEE, W. H. K., and RALEIGH, C. B. (1969), *Fault-plane Solution of the Koyna (India) Earthquake*,
Nature *223*, 172–173.

MCKENZIE, D. (1984), *A Possible Mechanism for Epirogenic Uplift*, Nature *307*, 616–618.

MERRITTS, D., and HESTERBERG, T. (1994), *Stream Network and Long-term Surface Uplift in the New
Madrid Seismic Zone*, Science *285*, 1081–1084.

MISHRA, D. C. (1989), *On Deciphering the Two Scales of the Regional Bouguer Anomaly of the Deccan
Trap and Crust-mantle Inhomogeneities*, J. Geol. Soc. India *33* (1), 48–54.

MORRISON, J. (1994), *Meteoric Water-rock Interaction in the Lower Plate of the Whipple Mountain
Metamorphic Core Complex, California*, J. Metamorphic Geol. *12*, 827–840.

MUIR-WOOD, R., and KING, G. C. P. (1993), *Hydological Signatures of Earthquake Strain*, J. Geophys.
Res. *98*, 22035–22068.

NEGI, J. G., PANDEY, O. P., and AGRAWAL, P. K. (1986), *Supermobility of Hot Indian Lithosphere*,
Tectonophys. *131*, 146–150.

NEGI, J. G., AGRAWAL, P. K., SINGH, A. P., and PANDEY, O. P. (1992), *Bombay Gravity High and
Eruption of Deccan Flood Basalt (India) from a Shallow Secondary Plume*, Tectonophys. *206*, 341–350.

NESBITT, B. E., and MUEHLENBACK, K. (1991), *Stable Isotopic Constraints on the Nature of the
Syntectonic Fluid Regime of the Canadian Cordillera*, Geophys. Res. Lett. *18*, 963–966.

PATIL, D. N., BHOSALE, V. N., GUHA, S. K., and POWAR, K. B. (1986), *Reservoir-induced Seismicity
in the Vicinity of Lake Bhatsa, Maharashtra, India*, Phys. Earth Planet. Int. *44*, 73–81.

QURESHY, M. N. (1981), *Gravity anomalies, isostasy and crust mantle relations in the Deccan Trap and
contiguous regions, India*. In *Deccan Volcanism and Related Basalt Provinces in Other Parts of the
World* (eds. Subba Rao, K. V., and Sukheswala, R. N.), Memoir Geol. Soc. India *3*, 184–197.

RADHAKRISHNA, B. P. (1993), *Neogene Uplift and Geomorphic Rejuvenation of the Indian Peninsula*, Current Sci. *64*, 787–793.

RAO, R. U. M., and JESSOP, A. M. (1975), *A Comparison of the Thermal Characters of Shields*, Can. J. Earth Sci. *12*, 347–360.

RASTOGI, B. K., and TALWANI, P. (1980), *Relocation of Koyna Earthquakes*, Bull. Seismol. Soc. Am. *70*, 1849–1868.

RAVAL, U. (1990), *A Rheological Waveguide as a Seat of Rifts and Basin Tectonics: In the Indo-Soviet Symposium on Rifted Basins and Aulacogens, Geological and Geochemical Approach*, Proc. of the Indo-Soviet Symp. on Rifted Basin and Aulacogens: Sedimentation, Crustal Evolution and Mineralization, January 22–30, Aligarh (eds. CASSHYAP, S. M. *et al.*) Gyanodaya Prakashan, Nainital, India, pp. 47–72.

RAVAL, U. (1993a), *The Regional Tectonothermal Setting of Killari (Latur) Area*. Proc. of the 30th Annual Convention and Seminar of IGU, Hyderabad, 21–23 Dec. 1993, 169–180.

RAVAL, U. (1993b), *On Certain Large-scale Gravity Field Patterns over the Subcontinent*. Proc. of the 30th Annual Convention and Seminar of IGU, Hyderabad, 21–23 Dec. 1993, 153–168.

RAVAL, U. (1993c), *Fast Movement of the Indian Plate, Buoyancy due to Plume and Warm Asthenosphere*, Proc. 29th Annual Convention and Seminar on Perspectives and Prospects in Geosciences towards the New Industrial Policy, 22–24 Feb. 1993 IGU, Hyderabad, pp. 104–114.

RAVAL, U., and VEERASWAMY, K. (1991), *Is Madagascar a Link between Southern and Northwestern Mobile Belts of India? Presented at the 28th Annual Convention of IGU, Hyderabad, Dec. 17–19, 1991*.

RICE, J. R., and CLEARY, M. P. (1976), *Some Basic Stress Diffusion Solutions for Fluid Saturated Elastic Porous Media with Compressible Constituents*, Rev. Geophysik *14*, 227–241.

SANDIFORD, M., and POWELL, R. (1986), *Deep Crustal Metamorphism during Continental Extension: Modern and Ancient Examples*, Earth. Planet. Sci. Lett. *79*, 151–158.

SARMA, S. V. S., VIRUPAKSHI, G., HARINARAYANA, T., MURTHY, D. N., PRABHAKAR, S., RAO, E., VEERASWAMY, K., MADHUSUDHAN RAO, SARMA, M. V. C., and GUPTA, K. R. B. (1994), *A Wide-band Magnetotelluric Study of the Latur Earthquake Region, Maharashtra, India*, Mem. Geol. Soc. India *35*, 101–118.

SHANKLAND, T. J., and ANDERS, M. E. (1983), *Electrical Conductivity, Temperature and Fluids in the Lower Crust*, J. Geophys. Res. *88*, 9475–9484.

SIBSON, R. (1987), *Earthquake Rupturing as a Mineralizing Agent in Hydrothermal Systems*, Geology *15*, 701–704.

SINGH, R. N., and NEGI, J. G. (1982), *High Moho Temperature in the Indian Shield*, Tectonophys. *82*, 299–306.

SINGH, D. D., RASTOGI, B. K., and GUPTA, H. K. (1975), *Surface-wave Radiation Pattern and Source Parameters of the Koyna Earthquake of December 10, 1967*, Bull. Seismol. Soc. Am. *65*, 711–731.

SINGH, S., AGRAWAL, P. N., and ARYA, A. S. (1976), *Filling of Ramganga Reservoir Kalagarh, U. P., India and its Possible Influence on Seismic Activity*, Bull. Seismol. Soc. Am. *66*, 1727–1731.

SIVARAMAN, T. V., and RAVAL, U. (1992), *U-Pb Isotopic Study of Certain Granites in the Rajasthan: An Evidence for the Late Cretaceous Thermal Remobilization*, J. Geol. Soc. of India (in press).

STEIN, M., and HOFMANN, A. W. (1994), *Mantle Plumes and Episodic Crustal Growth*, Nature *372*, 63–68.

STUART-ALEXANDER, D. E., and MARK, R. K. (1976), *Impoundment-induced Seismicity Associated with Large Reservoirs*, U. S. Geol. Surv., Open File Rep. 76–770.

WENZEL, F., and SANDMEIER. (1992), *Geophysical Evidence for Fluids in the Crust Beneath the Black Forest, S. W. Germany*, Earth Sci. Rev. *32*, 61–75.

YARDLEY, BRUCE W. D., and VALLEY, W. J. (1994), *How Wet is the Earth's Crust?* Nature *371*, 205–206.

(Received September 27, 1994, revised/accepted March 22, 1995)

PAGEOPH, Vol. 145, No. 1 (1995)

0033–4553/95/010193–15$1.50 + 0.20/0

Tomographic Inversion for the Three-dimensional Seismic Velocity Structure of the Aswan Region, Egypt

MOHAMED AWAD[1] and MEGUME MIZOUE[2]

Abstract—The Thurber iterative simultaneous inversion program is used to determine the three-dimensional *P*-wave velocity structure in the Aswan seismic region of Egypt. The tomographic inversion presented in this study is based on 1131 *P*-phase observations at 13 stations from 89 local earthquakes, all of which occurred within the Kalabsha fault zone. The assumed initial velocity model is that deduced from local explosion experiments. The results indicate that the Aswan region is characterized by a heterogeneous crust, consisting of a shallow, low-velocity zone and a deeper high-velocity anomaly. Seismic velocity structure within the shallow part demonstrates that the inferred change in velocity exists primarily across the east-west trending Kalabsha fault scarp, whereas the high-velocity zone is located south of this fault. Two well-resolved, low-velocity zones appear within the upper 6 km of the crust. The first coincides with a graben structure located between the Kalabsha and Seiyal faults and the second exists between the N–S Kurkur fault and the main axis of Lake Aswan. Both low-velocity zones occupy an area of approximately 30 × 40 km, located along the western bank of the lake. The most significant result of this study is that the location of the deeper, high-velocity anomaly coincides with the concentration of seismic activity in the lower crustal layer.

Key words: Velocity structure, inversion, earthquakes.

1. Introduction

The November 14, 1981, magnitude 5.7 Aswan earthquake, which occurred 70 km southwest of the Aswan High Dam (Fig. 1), has generated intense interest regarding the nature of the Aswan seismic zone. The primary focus of this interest is the relationship of the seismic activity to the Aswan Lake, which is the second-largest man-made reservoir on earth. Determination of the local seismic velocity structure is also being regarded as another priority, particularly its relationship to the tectonic setting and earthquake activity. This study represents the first attempt to determine the three-dimensional *P*-wave velocity structure for the Aswan

---

[1] National Research Institute of Astronomy and Geophysics, Helwan, Cairo, Egypt.

[2] Earthquake Research Institute, Tokyo University, 1-1 Yayoi 1-Chome, Bunkyo-ku, Tokyo 113, Japan.

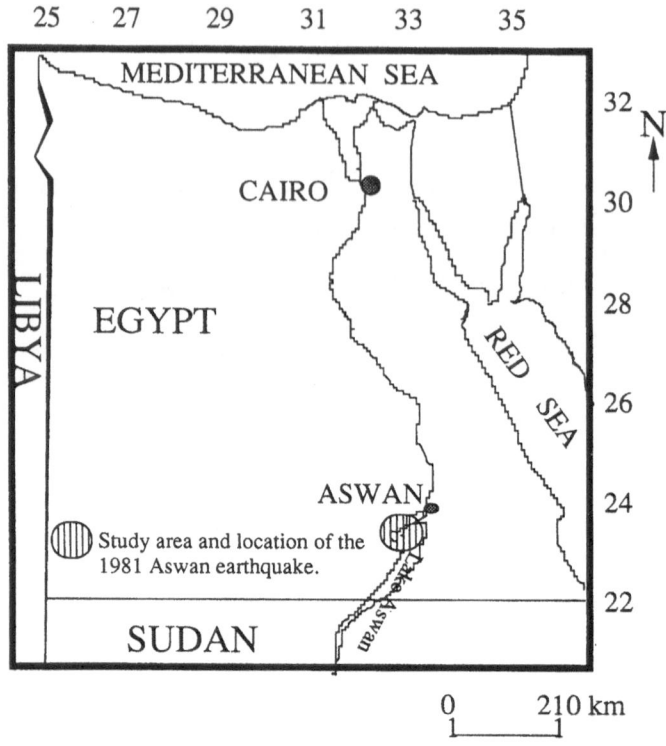

Figure 1
Study area.

region. The inferred crustal image is well constrained and clearly correlated with the local geological structure.

The November 14, 1981 earthquake occurred at a depth of 20 km (ISC Bulletin) along the Kalabsha fault zone, which is located beneath a large westward embayment of Lake Aswan (KEBEASY et al., 1982). This earthquake was followed by an extended sequence of small magnitude earthquakes ($M < 5$) and is characterized by right-lateral, strike-slip motion, consistent with the geologically inferred slip motion of the Kalabsha fault (KEBEASY et al., 1982, 1987; SIMPSON et al., 1987, 1992; BULOS et al., 1987; AWAD and MIZOUE, 1994).

The three-dimensional velocity inversion performed in this study is based upon local earthquake arrival-time residuals observed, using a local seismic network (Fig. 2). An earlier seismic refraction experiment in the Aswan area was performed in 1986 by the National Research Institute of Astronomy and Geophysics. The model deduced from this experiment indicates relatively low seismic velocities within the upper part of the granitic layer (KEBEASY et al., 1992) (Table 1), which has been

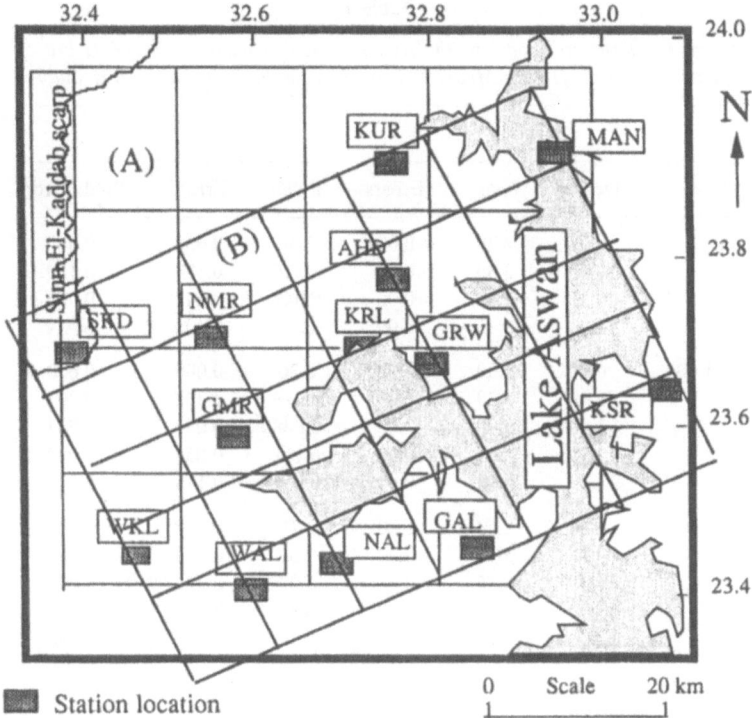

(A) Grid point configuration associated with the highest resolution values.

(B) Representative example of a trial grid point configuration, which is characterized by lower resolution values.

Figure 2
Grid point distribution around the Aswan network and the study region.

attributed to the effects of weathering. This velocity structure is adopted as an initial model for the three-dimensional inversion presented in our study.

The geological setting of the Aswan region is summarized in reports by ISSAWI (1968, 1971, 1978) and EL-SHAZLY (1977). The region is dominated by a succession of tectonic, erosional and igneous events that produced the present complex of granite and other rock types exposed at the surface within the Nubian sandstone plain. The tectonic characteristics indicate that this area is dominated by two sets of fault systems and regional uplift. The faults trend approximately in the E–W and N–S directions (Fig. 3). The Kalabsha fault trends E–W and cuts a limestone formation, known as Gebel Marawa, which is a remnant outcrop of a plateau that has receded to the west and now forms the Sinn El-Kaddab scarp. The Sinn El-Kaddab rises abruptly above a broad plain of Nubian sandstone.

Table 1

Local crustal structure model in the Aswan region, Models A, B and C are after KEBEASY *et al. (1992); Model D is used as the initial model in this study*

	Strating model			Calculated model		
	Depth	P-vel.	Errors	P-vel.	Errors	Pred. errors
Model (A)	0.00	2.30	0.00	2.30	0.000	0.000
	0.40	5.30	0.00	5.516	0.120	0.160
	4.30	6.05	0.00	5.947	0.049	0.066
	11.00	6.35	0.00	6.35	NAN	NAN
Model (B)	0.00	2.30	0.00	2.30	0.000	0.000
	0.40	5.30	0.00	5.218	0.392	0.207
	1.70	5.90	0.00	6.008	0.082	0.043
	4.30	6.05	0.00	6.111	0.056	0.030
	13.20	6.32	0.00	6.554	0.143	0.075
Model (C)	0.00	2.10	0.00	2.10	0.000	0.000
	0.30	5.30	0.00	5.67	0.043	0.052
	4.30	6.35	0.00	6.03	0.026	0.031
	13.00	6.80	0.00	6.32	0.046	0.055
	22.00	7.10	0.00	6.90	0.082	0.098
Model (D)	1.00	5.30				
	6.00	6.30				
	14.00	6.80				
	22.00	7.10				

2. Methodology

The tomographic inversion method used is that developed by THURBER (1981, 1983), which is based upon the iterative inversion of P-wave arrival-time data, for the simultaneous determination of three-dimensional crustal velocity structure and hypocentral parameters. This method uses the arrival-time data of earthquakes located within the target region and recorded at local stations. The target region is divided into small blocks, within which the slowness perturbations are calculated on the basis of an interpolation function (THURBER, 1983, 1986). Preliminary estimates of the earthquake parameters (hypocenter and origin time) are obtained by applying the initial velocity model, allowing variation in depth only, and the ray paths are calculated from the source to each of the observing stations. This inversion method involves three steps: (1) Velocity model parameterization; (2) parameter separation; and (3) approximate ray tracing calculation (ART) (THURBER, 1981, 1983, 1986). Parameter separation is necessary to minimize computer storage requirements. Background information on simultaneous inversion can be found in AKI and LEE (1976), SPENCER and GUBBINS (1980), PAVLIS and

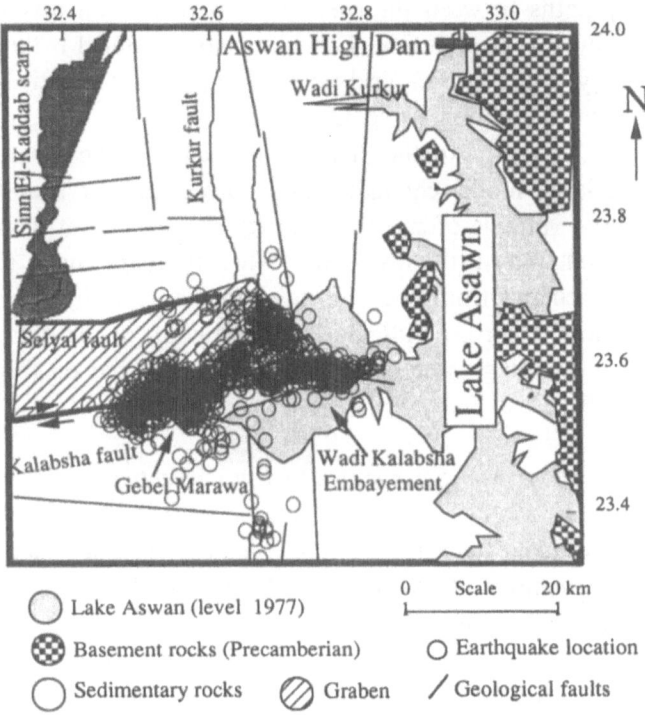

Figure 3
Geological and seismicity map of study area.

BOOKER (1980) and THURBER (1983, 1984, 1986). In the approximate ray tracing method (ART), a set of smooth curves connecting the source and the receiver is constructed as initial estimates of the ray paths, and the travel time along each curve is calculated. Arcs of varying radii are examined and the dip of the plane containing the arcs is varied systematically. The path characterized by the shortest travel time is selected as the true ray path. Travel-time estimates by this method agree quite well with the 'true' ray path travel times calculated, using a three-dimensional ray tracer for the short distance domain (<50 km), having a maximum standard deviation of 0.02 s (THURBER, 1983). This approximate method is limited, however, in the sense that the path curvature is constant along a given curve and each curve lies within a single plane. Pseudo-bending (an iterative approach) is introduced to perturb the (ART) ray path so that it satisfies the constraint that the curvature of the true ray path is always perpendicular to the normal component of the local velocity gradient (e.g., UM and THURBER, 1987). This enables a given ray to have variable curvature and to deviate from a single plane. Velocity values are assigned at fixed points, which define the grids in the three-dimensional domain. Seismic velocity values and their partial derivatives are assumed to vary in a linear manner along ray paths between the specified grid points, an assumption which produces a gradually varying solution. The final velocity model is then determined by assigning velocity values at each grid point. The program iteratively determines a damped least-squares solution using LS decomposition. A damping parameter, defined by the user, is added to the diagonal elements of the separated medium matrix to prevent large model changes that would occur for near-zero singular values. The damping value is often chosen to equal the ratio of data variance to model variance (EBERHART-PHILLIPS, 1986). When the damping is too small, the predicted velocity oscillates from one grid point to the next, and large changes in velocity result without a correspondingly large reduction in the data variance. In the present study a damping value of 0.5 was assigned, based on this approach.

3. Aswan Seismicity and Data Selection

Most of the earthquakes constituting the extended sequence that followed the Aswan event on November 14, 1981 were well recorded by the Aswan seismic network. This network consists of 13 field stations (Fig. 2) and is a typical narrow-band frequency modulated (FM) system (SIMPSON et al., 1987). Each field station is equipped with a single-component vertical seismometer, except those at the GRW and GMR locations, both of which have two three-component seismometers. The field stations are well distributed around the Kalabsha fault, which is assumed to be the source of the November 14, 1981 event. Since June 1982, there were only two recognized peaks in Aswan seismic activity. The first occurred in August 1982 within the deep part of the crust (i.e., at depths from 18 to 26 km), with the largest magnitude ($M = 4.9$) event occurring on August 20, 1982. The second swarm-type burst occurred at shallow depth (i.e., from 4 to 9 km) during June 1987. Aswan seismic activity is characterized by two distinct zones. The deeper seismic zone is located beneath Gebel Marawa along the Kalabsha fault (Fig. 8), whereas the shallow seismic zone is represented by activity along the eastern segment of the Kalabsha fault and the southernmost part of the Kurkur fault (Fig. 3). The epicentral distribution of earthquakes (Fig. 3), as well as the focal mechanism solutions (AWAD and MIZOUE, this issue) indicate that this seismicity is well correlated with the inferred tectonic setting. Furthermore, the presence of a deep seismic zone in this region presents an excellent opportunity to investigate the seismic velocity structure within most of the crustal layer.

For the purpose of this study, the following criteria were used for choosing events from the Aswan data base: (1) Location within the target volume; (2) number of picks greater than 10 for each event; (3) accurately synchronized events recorded on analog tape; and (4) a clear record of the P and S phases.

The present study was facilitated by a seismic refraction experiment performed within the Aswan region that provides a preliminary seismic velocity model (Table

Table 2

Station corrections for the Aswan seismic stations. C_p is the P-arrival time correction and C_s is the S-arrival time correction

Station code	C_p (s)	C_s (s)	Elevation (m)	Bedrock type
AHD	0.211	−0.171	222	sandstone
KSR	0.218	−0.578	260	sandstone
GAL	0.105	−0.368	278	sandstone
SKD	0.107	−0.107	270	sandstone
NMR	0.135	−0.241	200	clay-sand
GRW	−0.118	0.313	260	hard rocks
WKL	−0.133	−0.560	203	clay-sand
WAL	−0.050	−0.279	197	clay-sand
NAL	−0.092	−0.341	150	sandstone
KRL	−0.177	−0.145	185	sandstone
KUR	−0.321	−0.564	190	sandstone
MAN	−0.392	−0.485	195	hard rocks

1) (KEBEASY et al., 1992). Because the seismograms of this experiment are not available, the present study is limited to use of the local earthquake data. Earthquake hypocentral parameters are determined using an inversion program developed by HIRATA and MATSU'RA (1987) and were associated with significant travel-time residuals. Because the residuals were very high, station correction (i.e., the mean residuals) is estimated by dividing the sum of the residuals by the number of events per station. These corrections (Table 2) can be related to the structural anomalies in the immediate vicinity of the stations because the seismometers are situated over topography of various geologic structures. Earthquake hypocentral parameters are redetermined after applying the station corrections. Relocated earthquakes that are characterized by travel-time residuals of less than 0.25 s and uncertainties in the epicentral location and focal depth of less than 1.5 km and 2.0 km, respectively, were selected for the seismic velocity inversion. The 89 events selected provide 1131 observations of P-wave arrival times. These earthquakes are characterized by different focal depths, as shown in the histogram in Figure 4.

4. Velocity Modeling

Several nodal configurations were tested to determine the optimal velocity model parameterization. A representative example of a trial grid point configuration is denoted "B" in Figure 2. Both the limited number of stations spanning a large area and the concentration of earthquakes within a narrow zone presented in the Aswan region, do not allow the fine-spacing modeling of grid points (e.g., 5 km),

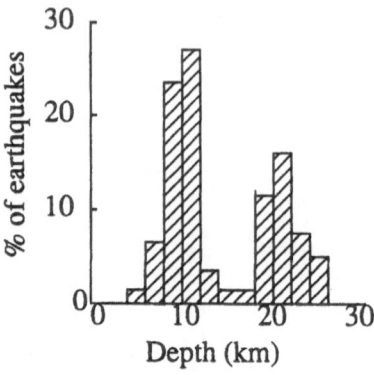

Figure 4
Histogram showing depth distribution of the hypocenters.

which has been tested. Selection of the optimal grid point configuration is facilitated by connecting the source-receiver locations. This test indicates that the variations on a scale finer than 15 km will be poorly defined and will require an increase in the number of properly located seismic stations, as well as the use of artificial sources of seismic waves. The configuration denoted "A" in Figure 2 is associated with the most pronounced diagonal element resolution. It contains five layers of constant seismic velocity, the upper boundaries of which occur at depths of −0.5, 1, 6, 14 and 22 km, with P-wave velocities of 2.8, 5.3, 6.3, 6.8 and 7.1 km/s, respectively. The grid point pattern of this model is aligned parallel to the regional geological structures in the Aswan region and covers an area of approximately 64×63 km in the E–W and N–S directions (Fig. 2), respectively. Grid points in the horizontal plane are located at the intersection of orthogonal planes of coordinates 1, 15, 30, 45 and 65 km in the E–W direction and 2, 16, 32, 48 and 65 km in the N–S direction. These coordinates are measured with respect to the point 23°22′N and 32°22′E. Intersections of these orthogonal planes with the horizontal surfaces of the five layers discussed above define the three-dimensional grid point distribution. Except for the northwestern part of the modeled space, where there are no observation sites, the grid points are generally located in regions of maximum ray coverage. In order to test the effect of variation of the depth distribution of grid points on the final velocity model, depths of the layers were reassigned at −1, 1, 6, 10 and 22 km. Although this model produced no significant change in seismic velocity values, it is characterized by slightly lower resolution values (of maximum 0.70). A small change in the initial seismic velocity model (e.g., 10%) produced no significant effect on velocity values (average 0.08 km/s), but produced a change in depths of the event foci within +2 km. The concentration of earthquakes within a narrow zone along the faults in this region limits the possibility of quantifying the influence of event distribution. A limited test, however, was performed by using two subsets of events: one which is restricted to those along the Kalabsha fault and one which contains additional events from the southern part of the area (Fig. 3). Although both subsets of events produced similar seismic velocity perturbations, the rigorousness of this test is obviously limited by the concentration of events within a narrow zone along the faults in this region.

Collectively, these tests demonstrate that the final seismic velocity model is stable with respect to node spacing, the available earthquake data, and stations locations. The influence of event distribution, however, is poorly characterized. An increase in the number of properly seismic stations, along with artificially generated seismic wave experiments, would significantly improve subsequent studies of this nature.

5. Results

Deviation of the final velocity solution from the initial model at depths of 1, 6, 14 and 22 km is shown in Figure 5. This deviation achieved a value of 0.55 km/s at some nodes. The corresponding values in the resolution matrix are contoured in Figure 6. The diagonal resolution element values range from 0.0 to 0.92. Higher diagonal resolution element values indicate that the velocity solution is primarily influenced by the velocity in the space immediately surrounding the grid point. The low resolution values associated with several nodes in the final solution may be attributed to a lack of stations within these areas and/or the concentration of deeper seismicity within a small zone. The largest values in the resolution matrix are near the center of the region of interest, where a relatively dense distribution of both earthquakes and stations exists. The final velocity model is also characterized by small values of calculation errors, which are contoured in Figure 7. The calculation error represents the ratio of the sum of square residuals of the final velocity model to that of the previous iteration multiplied by the standard error at each grid point, and is indicated as a percentage value. The final model is characterized by a standard deviation of 0.07, an rms of 0.046 and a sum of squared residuals of 1.7. The velocity variations are, therefore, significant with respect to the uncertainties. It is important to note, however, that errors may be produced by: (1) The spatial definition of a discrete grid of velocity values and distribution of data; (2) insignificant model perturbation within the area of low resolution, which, in this case, reflects the original one-dimensional model; and (3) error in calculating the velocity perturbations which for P-wave velocity solutions ranged from 0.2 to 0.39 km/s.

6. Discussion and Conclusion

The prominent velocity contrast in the shallow portion of the three-dimensional model exhibits a remarkable correlation with local surface geology. Analysis of all

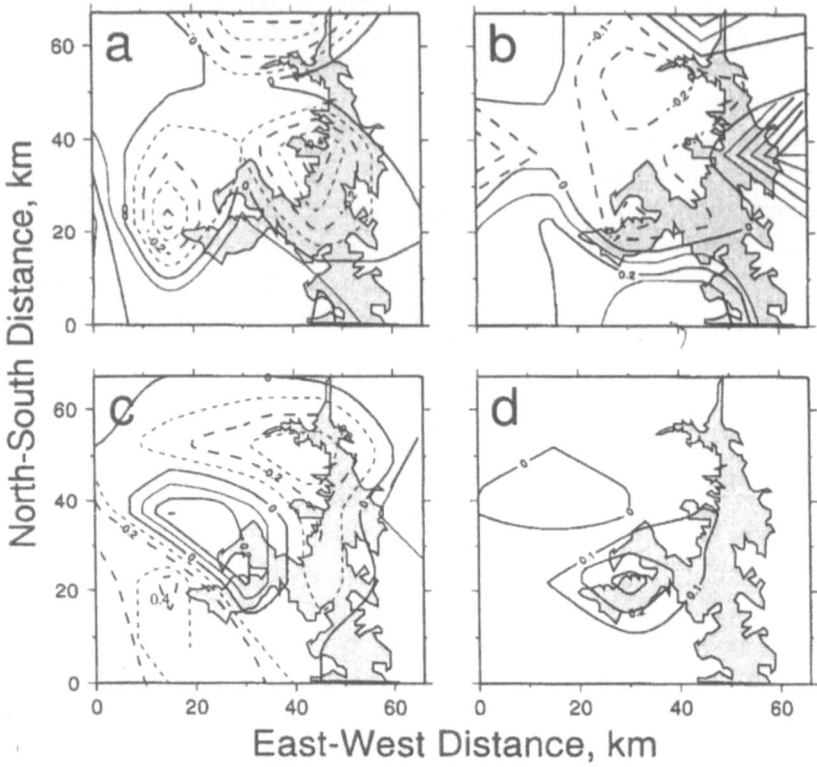

Figure 5
Contour maps in a, b, c and d showing the deviation of final P-wave velocity structure from the initial model of 5.3, 6.3, 6.8 and 7.1 km/s at depths of 1, 6, 14 and 22 km, respectively. Broken lines denote negative deviation.

plots indicates that the velocity contrast exists at depths to 6 km, apparent in the velocity deviations illustrated in Figure 5. ISSAWI (1978) has mapped the two fault systems within the study region (Fig. 3). The Kalabsha and Seiyal faults, which trend in an approximately E–W direction, constitute a graben structure occupied by the Kalabsha embayment. The change in velocity within the shallow crust evidently occurs directly across the Kalabsha fault zone. The lower velocities coincide with the down-thrown blocks of the faults, whereas the higher velocities are located along the footwall of the Kalabsha fault and are terminated by the main axis of Aswan Lake. This contrast may be due to the elevation of higher velocity basement rocks in the footwall relative to the graben block, or, possibly, a reduction of seismic velocities associated with the presence of fluids within the graben block and fault zone. Because the velocity is lowest north of the Kalabsha fault, which is a seismically active region (Fig. 3), it may also be possible that the velocity reduction in the graben block is partially attributable to the presence of

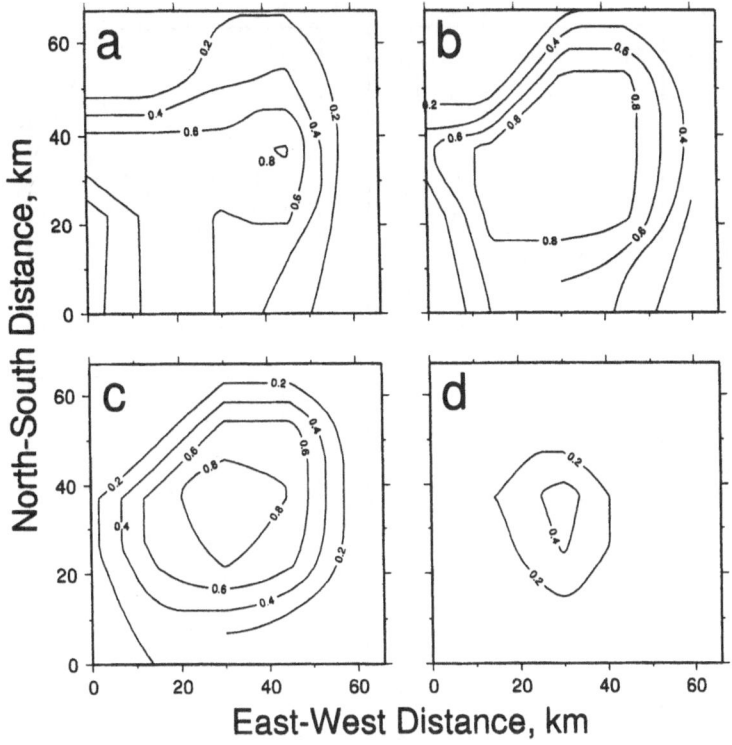

Figure 6
Contours of the *P*-wave resolution values from the resolution matrix. a) 1 km, b) 6 km, c) 14 km, d) 22 km.

fluids and/or microcracks. The seismic activity in this region is shown with respect to depth distribution in the cross sections of Figure 8. The values of *P*-wave velocity in the high-resolution zone are also presented in these cross sections by the contour lines. In the N–S trending cross section, the shallow seismicity is seen in the low-velocity zone, which is characterized by a Vp of 4.8 km/s at 1 km depth and 6.07 km/s at 6 km depth. Note, however, that the same depth levels in the southern part are characterized by Vp values of 5.44 and 6.66 km/s, respectively. The shallow seismicity of this region exhibits a typical swarm pattern, a small b value of the Gutenberg-Richter relationship, and a very large number of small magnitude earthquakes, which demonstrate the presence of microcracks (AWAD and MIZOUE, 1994).

In the deeper crust, at a depth of 14 km (Fig. 5), a high-velocity anomaly is evident. The *P*-wave velocity at this depth reaches an average of 7.03 km/s, which is remarkably high with respect to the surrounding values (average 6.5 km/s). The cross sections in Figure 8 indicate that the deeper seismic zone, which is character-ized by a foreshock-main shock-aftershock sequence and a normal *b* value (i.e.,

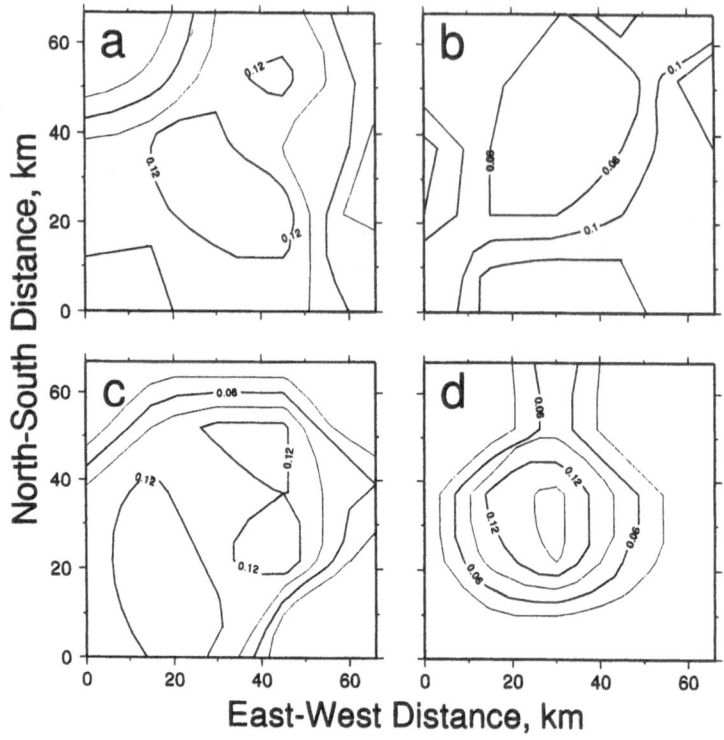

Figure 7
Contours of the calculation errors associated with the final velocity model. a) 1 km, b) 6 km, c) 14 km,
d) 22 km.

0.99) (AWAD and MIZOUE, 1994), is located within the high-velocity zone. More-over, these deeper earthquakes are more concentrated than the scattered activity observed in the shallow seismic zone. The Aswan region is located at the interface between different tectonic blocks, the Arabo-Nubian massif and the Pan African shields (EL-SHAZLY, 1977). DIXON and GOLOMBEK (1988) located the contact between these two blocks through isotopic analysis of the Aswan granite. This area is also characterized by regional uplift (ISSAWI, 1971, 1978). The velocity depth distribution (Fig. 8) shows a lateral variation with an increase in velocity toward the southwest. The change in seismic velocity may therefore be related to litholog-ical variations between the crustal blocks forming this region, as well as the presence of a deeper seismic zone.

Changes in seismic velocity have been related to the variation of pressure and/or temperature within the medium (RALEIGH and EVERNDEN, 1981), with zones of high pressure and/or high temperature corresponding to reduced seismic velocity. The presence of both low- and high-velocity crustal intrusives has been correlated with the results of inverted three-dimensional crustal structure (e.g., CROSSON,

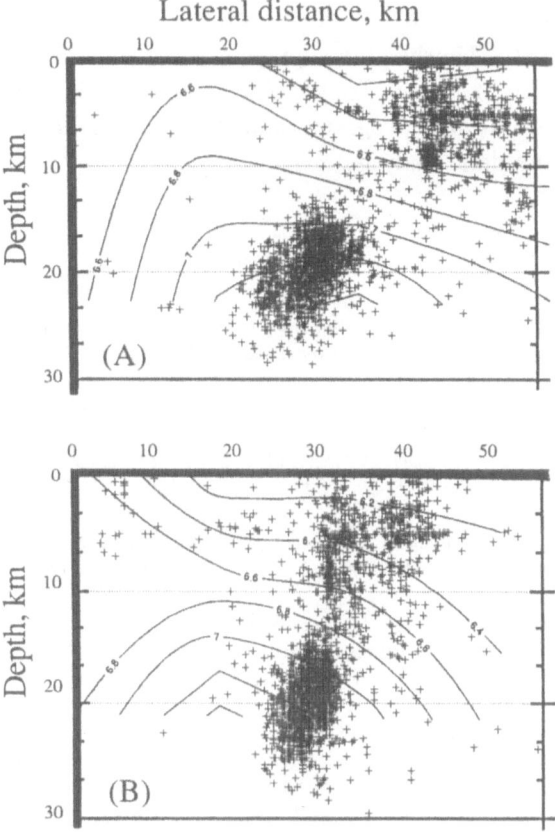

Figure 8

Correlation between seismicity and P-wave velocity structure in the Aswan region, (A) and (B) are the E–W and N–S cross sections, respectively. + earthquake foci; values of the contour lines are the P-wave velocity (km/s).

1976; EBERHART-PHILLIPS, 1986; FOULGER and TOOMEY, 1989). Other geophysical data, however, are required to confirm this interpretation in the Aswan region. Detailed investigation of the local crustal structure, using a dense seismic network, is the next logical step.

The three-dimensional P-wave velocity analysis presented in this study demonstrates that the Aswan region is characterized by a heterogeneous crust, with a low-velocity zone within the shallow part and a deeper high-velocity anomaly. In this region, the shallow seismic activity occurred within the low-velocity zone, and the deeper seismic activity is concentrated within the deeper high-velocity anomaly.

Acknowledgements

We thank Prof. C. Thurber for supplying a copy of his 3-D program to the Earthquake Research Institute, Tokyo University, Dr. S. Mueller for assistance with the GMT contouring program, and Dr. T. Miyatake for his generous assistance during the data analysis.

REFERENCES

AKI, K., and LEE, W. H. K. (1976), *Determination of Three-dimensional Velocity Anomalies under a Seismic Array Using First P-arrival Times from Local Earthquakes. 1. A Homogeneous Initial Model,* J. Geophys. Res. *81,* 4381–4399.

AWAD, M., and MIZOUE, M. (1994), *Seismicity Features in the Active Region beneath Aswan High Dam Lake, Egypt,* Abstract for the IASPEI94, 27 General Assembly.

BOULOS, F. K., MORGAN, P., and TOPPOZADA, T. (1987), *Microearthquake Studies in Egypt Carried out by the Geological Survey of Egypt,* J. Geodynamics 7, 227–249.

CROSSON, R. S. (1976), *Crustal Structure Modeling of Earthquake Data, 1, Simultaneous Least Squares Estimation of Hypocenter and Velocity Parameters,* J. Geophys. Res. *81,* 3036–3046.

DIXON, T. H., and GOLOMBEK, M. P. (1988), *Late Precambrian Crustal Accretion Rates in Northeast Africa and Arabia,* Geology 16, 991–994.

EBERHART-PHILLIPS, D. (1986), *Three-dimensional Velocity Structure in Northern California Coast Ranges from Inversion of Local Earthquake Arrival Times,* Bull Seismol. Soc. Am. 76, 1025–1052.

EL SHAZLY, E. M., *The Geology of the Egyptian Region. The Ocean Basin and Margins 4A* (Plenum Publishing Corporation 1977) pp. 379–444.

FOULGER, G. R., and TOOMEY, D. R. (1989), *Tomographic Inversion of Local Earthquake Data from the Hengill-Grensdalur Central Volcano Complex, Iceland,* J. Geophys. Res. 94, B12, 17, 497–517, 510.

HIRATA, N., and MATSU'RA, M. (1987), *Maximum-likelihood Estimation of Hypocenter with Origin Time Eliminated Using Nonlinear Inversion Technique,* Phys. Earth and Planet. Int. 47, 50–61.

ISSAWI, B. (1968), *The Geology of Kurkur-Dungul Area,* General Egyptian Organization for Geological Research and Mining, Geological Survey, Cairo, Paper No. 46, 101 pp.

ISSAWI, B. (1971), *Geology of the Darb el-Arbain, Western Desert,* Annals of the Geological Survey of Egypt.

ISSAWI, B. (1978), *Geology of the Nubia West Area. Western Desert,* Annals of the Geological Survey of Egypt.

KEBEASY, R. M., MAAMOUN, M., and IBRAHIM, E. M. (1982), *Aswan Lake Induced Earthquakes,* Bull. Inter. Inst. of Seismol. and Earthq. Eng., Tokyo, *19.*

KEBEASY, R. M., MAAMOUN, M., IBRAHIM, E. M., MEGAHED, A., SIMPSON, D. W., and LEITH, W. S. (1987), *Earthquake Studies at Aswan Reservoir,* J. Geodyn. 7, 173–193.

KEBEASY, R. M., BAYOUMI, A. I., and GHARIB, A. A. (1992), *Crustal Structure Modelling for the Northern Part of the Aswan Lake Area Using Seismic Waves Generated by Explosions and Local Earthquakes,* J. Geodyn. *14,* 1–24.

PAVLIS, G. L., and BOOKER, J. R. (1980), *The Mixed Discrete-continuous Inverse Problem: Application to the Simultaneous Determination of Earthquake Hypocenters and Velocity Structure,* J. Geophys. Res. *85,* 4801–4810.

RALEIGH, B., and EVERNDEN, J. (1981), *Case for Low Deviatoric Stress in the Lithosphere. Mechanical Behavior of Crustal Rocks,* The Handin Volume, Geophysical Monograph *24,* 173–186.

SIMPSON, D. W., KEBEASY, R. M., NICHOLSON, C., MAAMOUN, M., ALBERT, R. N. H., IBRAHIM, E. M., MEGAHED, A., GHARIB, A., and HUSSAIN, A. (1987), *Aswan Telemetered Seismograph Network,* J. Geodyn. 7, 195–203.

SIMPSON, D. W., GHARIB, A. A., and KEBEASY, R. M., *Induced seismicity and changes in water level at Aswan reservoir, Egypt,* In *Induced Seismicity* (ed. Knoll, P.) (Central Institute for Physics of the Earth, Potsdam 1992) pp. 331–344.

Spencer, C., and Gubbins, D. (1980), *Travel-time Inversion for Simultaneous Earthquake Location and Velocity Structure Determination in Laterally Varying Media*, Geophys. J. R. Astron. Soc. *63*, 95–116.

Thurber, C. H. (1981), *Earth Structure and Earthquake Locations in the Coyote Lake Area, Central California*, Ph. D. Thesis, MIT, MA, U.S.A.

Thurber, C. H. (1983), *Earthquake Locations and Three-dimensional Crustal Structure in the Coyote Lake Area, Central California*, J. Geophys. Res. *88*, 8226–8236.

Thurber, C. H. (1988), *Analysis Methods for Kinematic Data from Local Earthquakes*, Rev. Geophys. *24* (4), 793–805.

Um, J., and Thurber, C. H. (1987), *A Fast Algorithm for Two-part Seismic Ray Tracing*, Bull. Seismol. Soc. Am. *77*, 972–986.

(Received March 23, 1994, revised/accepted August 31, 1994)

PAGEOPH, Vol. 145, No. 1 (1995)

0033–4553/95/010209–09$1.50 + 0.20/0

Reservoir Associated Characteristics Using Deterministic Chaos in Aswan, Nurek and Koyna Reservoirs

H. N. Srivastava,[1] S. N. Bhattacharya,[2] K. C. Sinha Ray,[1] S. M. Mahmoud,[3] and S. Yunga[4]

Abstract—Nurek, Aswan and Koyna reservoirs were affected by moderate earthquakes with continuing seismic activity. Microearthquake data recorded through local networks have been used to determine the strange attractor dimensions, using deterministic chaos which were found as 7,2, 3.8 and 4.8, respectively. This would imply that while 8 parameters are needed to model earthquakes near Nurek reservoir, only 4 to 5 parameters are needed for the Aswan and Koyna regions. The differences in the strange attractor dimension suggest them to be a measure of seismotectonics around such reservoirs.

Key words: Earthquakes, strange attractor, deterministic chaos, reservoir.

Introduction

Reservoir associated seismicity has necessitated the attention of geoscientists to evolve new approaches to understand the phenomenon. Significant contributions have nevertheless been made largely based on statistical methods (Gupta, 1992; Simpson, 1986; Srivastava *et al.*, 1991 and others). Using a dynamical approach, Srivastava *et al.* (1994) found that a strange attractor exists in the Koyna region and at least 5 parameters are needed for the earthquake predictability in the region. A question arises whether seismic activity around other reservoirs like Aswan and Nurek is characterized by a strange attractor. Also it is of interest to understand the number of parameters needed for earthquake predictability in different tectonic regions. The objectives of this paper are to examine these aspects of the Aswan and Nurek reservoirs and compare the results with those of the Koyna region.

---

[1] Meteorological Office, Pune-411005, India.
[2] Meteorological Office, New Delhi-110003, India.
[3] National Research Institute of Astronomy and Geophysics, (NRIAG), Helwan, Egypt.
[4] Institute of the Physics of Earth, Moscow, Russia.

Data

i) *Nurek reservoir.* Data of 22,000 earthquakes from 1976 to 1987 have been considered in this study which is based upon a radio telemetered network of 10 stations around the reservoir.

ii) *Aswan reservoir.* Microearthquake data for the period from 1982 to December 1990, based on a telemetered seismic array, were used for this study.

For tectonics around the Aswan and Nurek reservoirs, a reference may be made to KEBEASY *et al.* (1987) and SIMPSON and NEGMATULLAEV (1981).

Methodology

The method of GRASSBERGER and PROCACCIA (1983) was applied to calculate the fractal dimension of the strange attractor. A time series of variable x was constructed using the frequency of earthquakes in a 2-days block.

We define a point X_i in m dimensional phase space whose coordinates are:

$$x(t), \ x(t + \tau), \ldots, x(t + (m - 1)\tau),$$

where τ is the delay time, which is an integral multiple of the sampling time (2 days).

A set of N points, i.e. $X_1, X_2, \ldots, X_N$ on an attractor embedded in a phase space of m dimension is obtained from the time series $x(t_1), \ldots, x(t_N)$. The difference $|X_i - X_j|$ from the $N - 1$ remaining points is computed.

The correlation function of the attractor $C_m(r)$ is given by

$$C_m(r) = \frac{1}{N^2} \sum_{\substack{i,j=1 \\ i \neq j}}^{N} H(r - |X_i - X_j|) \tag{1}$$

for embedding dimension m, where H is Heavyside function, i.e.,

$$H(x) = 0, \quad \text{if } x \leq 0$$

$$H(x) = 1, \quad \text{if } x > 0$$

and r is distance.

The dimensionality d of the attractor is related to $C_m(r)$ by the relation

$$C_m(r) = r^d \quad (\text{where } r \text{ is small})$$

$$\ln C_m(r) = d \ln(r). \tag{2}$$

Hence, the dimensionality d of the attractor is given by the slope of the $\ln C_m(r)$ versus $\ln (r)$.

$\ln C_m(r)$ is plotted against $\ln(r)$. The slope of the scaling region is obtained for various embedding dimensions (Fig. 1). As we increase the embedding dimension

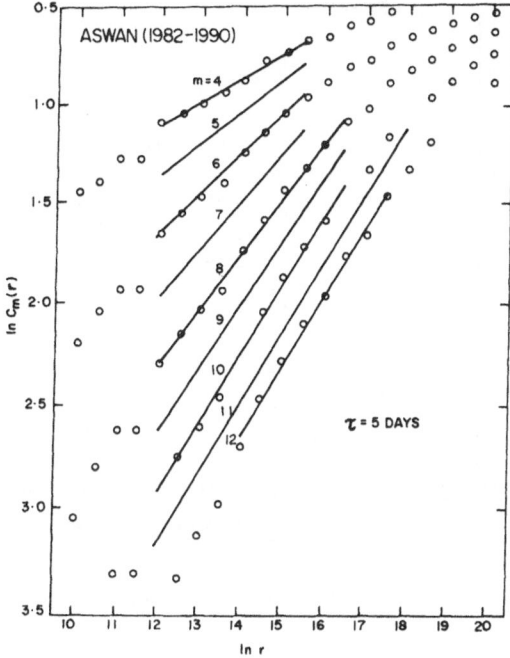

Figure 1(a)

ln $C_m(r)$ versus ln(r), $\tau = 5$ days, r being correlation length, for Aswan reservoir.

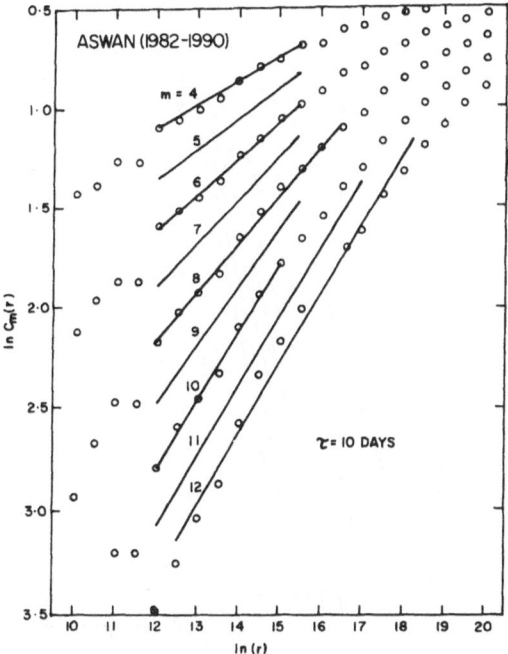

Figure 1(b)

ln $C_m(r)$ versus ln(r), $\tau = 10$ days, r being correlation length, for Aswan reservoir.

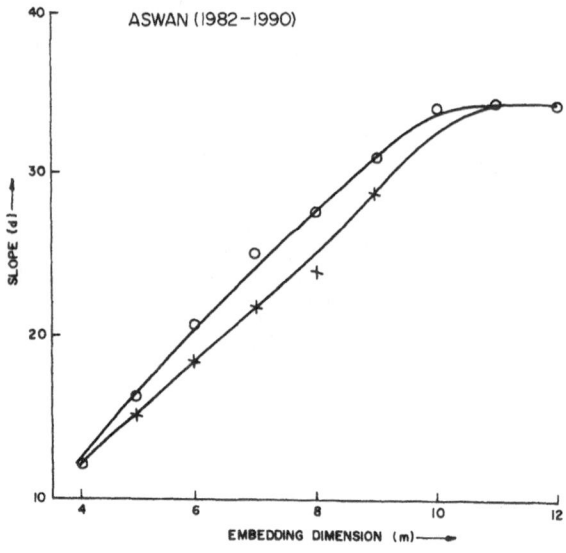

Figure 1(c)
Slope of the scaling region versus embedding dimension for the Aswan reservoir.

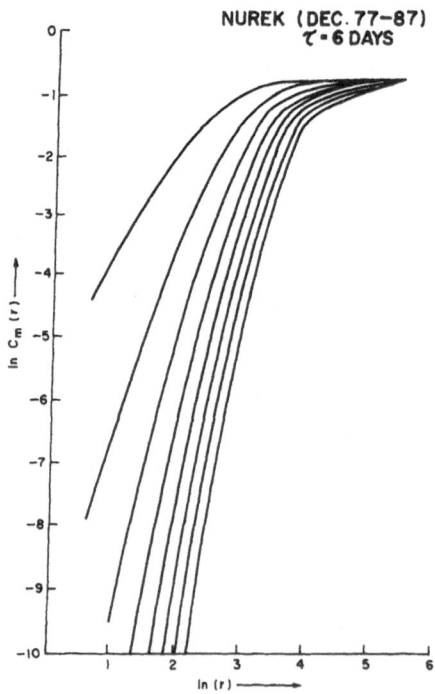

Figure 2(a)
ln $C_m(r)$ versus ln(r), $\tau = 6$ days, for Nurek reservoir.

NUREK (DEC. 77–87)

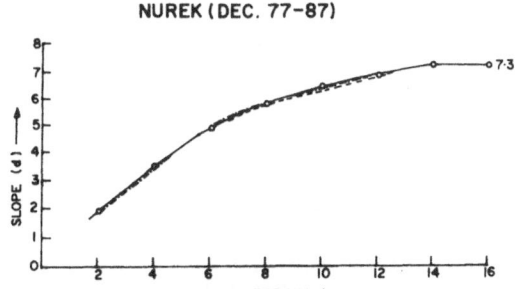

'Figure 2(b)
Slope of scaling region versus embedding dimension for the Nurek reservoir, $\tau = 2, 4, 6$ days.

m, the slope saturates to a limiting value which is considered as the fractal dimension of the strange attractor. The delay time τ is slowly increased until the same fractal dimension is obtained for two consecutive delay times. In the case of the Aswan reservoir, the fractal dimension obtained for the delay time $\tau = 5$ and 10 days remains the same (Fig. 1), which also lends support to the chaotic process around the reservoir as an example of deterministic chaos.

Figure 2 displays the plot of the embedding dimension in relation to the correlation integral for the Nurek reservoir. The strange attractor dimensions for the Aswan, Koyna and Nurek reservoirs are given in Table 1. Figure 3 illustrates the earthquake activity in three-dimensional phase space around the Aswan, Koyna and Nurek reservoirs.

It should be noted that care was taken to avoid spurious results from being obtained by keeping the number of earthquakes N such that RUELLE's criteria (1990), namely $2 \log_{10} N > D$, is satisfied where D is the fractal dimension. This criteria held good for the Aswan and Koyna reservoirs which maintained low fractal dimensions. For the Nurek reservoir which possessed a large fractal dimension, the number of earthquakes also satisfied the condition.

Table 1

Strange attractor dimensions and the largest earthquake in the Aswan, Koyna and Nurek reservoirs

No.	Region	Largest earthquake magnitude	Date	Strange attractor dimension
1.	Aswan	5.1	14 Nov 1981	3.4
2.	Koyna	6.5	11 Dec 1967	4.4
3.	Nurek	4.6	06 Nov 1972	7.2

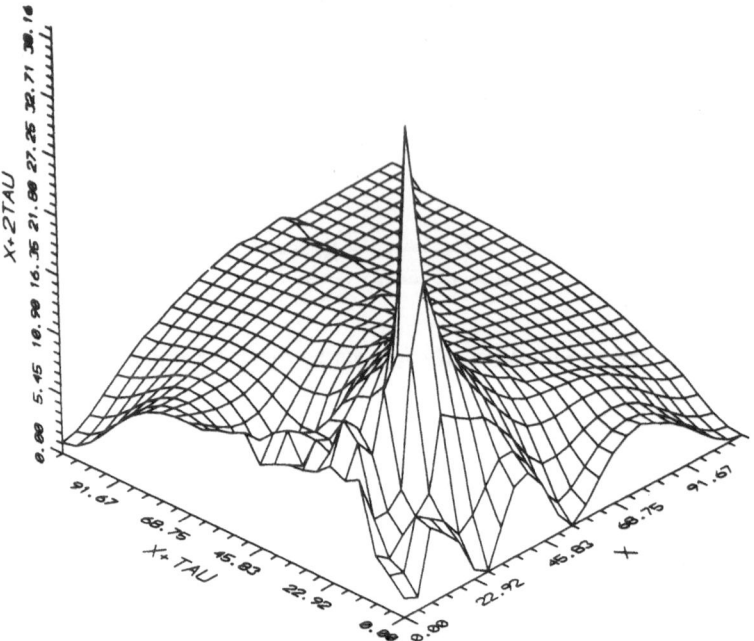

Figure 3(a)
Phase space diagram for the Aswan reservoir earthquakes.

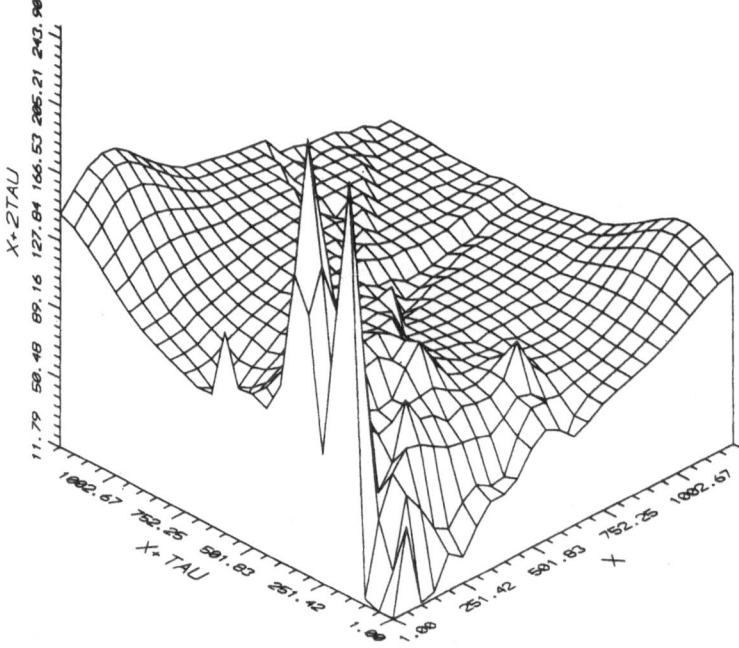

Figure 3(b)
Phase space diagram for the Koyna reservoir earthquakes.

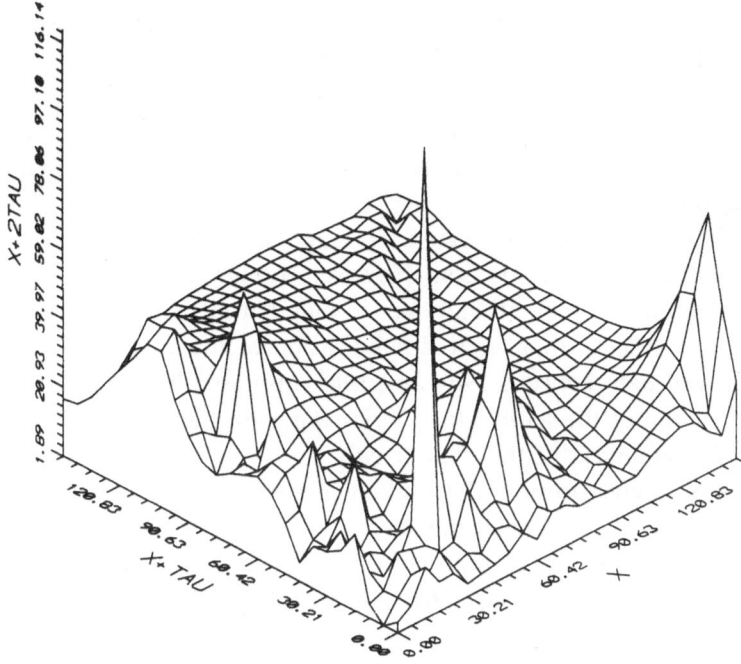

Figure 3(c)
Phase space diagram for the Nurek reservoir earthquakes.

Results and Discussion

It may be noted from Table 1 that since the dimensions are noninteger, the strange attractor exists around the Nurek and Aswan reservoirs. The results are similar to those reported for the Koyna reservoir (SRIVASTAVA *et al.*, 1994). In the Koyna region the fractal dimension of the strange attractor was 4.4, implying at least 5 parameters for earthquake predictability in the region. Considering that both earthquakes in the Aswan and Koyna regions occurred in a shield region remote from the active seismic plate boundary, the results are almost comparable. However, the largest earthquake in the Koyna region attained a magnitude of 6.5, with an associated longer fault length as compared to the Aswan reservoir region and therefore relatively larger parameters could be attributed to the differences in the tectonics.

On the other hand, the fractal dimension was 7.3 near the Nurek reservoir, implying at least 8 parameters for earthquake predictability, which is higher than that observed for the Aswan and Koyna regions which lie in stable tectonic regions. BHATTACHARYA and SRIVASTAVA (1992) found that the strange attractor in the Hindukush region had a fractal dimension of 6.9, which lies near the Indian-Eurasian plate. Thus, the higher fractal dimension in the Nurek Dam may be attributed to the complexity of tectonics (KEITH *et al.*, 1982). This is also supported by Figure 3 which manifests greater complexity of surface folding for the Nurek region as compared to the Koyna and Aswan reservoirs. Consequently, the differences in the strange attractor obtained for the earthquakes in the Koyna, Aswan and Nurek reservoirs suggest them to be a measure of seismotectonics of the respective regions.

For the Parkfield, California region, HOROWITZ (1989) reported that there is strong evidence for a chaotic process. However, JULIAN (1990) suggested further studies, based on more observational data in different tectonic regions. BELTRAMI and MARESCHAL (1993) found that the dimension of the attractor, based on the Parkfield network data between 1969 and 1987, appeared higher than 12 and the correlation function of the seismic time series was indistinguishable from that of a series of random numbers of the same length.

It may be of interest to note that in volcanic regions, SORNETTE *et al.* (1991) found the existence of low fractal dimensions of 2 to 4 in the dynamical systems underlying the time evolution in Mauna Loa and Kilavea, Hawaii. Keeping this in view, the findings of the low strange attractor near the Koyna and Aswan reservoirs could be attributed to different predictive parameters than those for volcanic eruptions.

Conclusions

The study has brought out the following:
i) A strange attractor exists in the Aswan, Koyna and Nurek reservoirs.
ii) The strange attractor dimension may be characteristic of the tectonics around the reservoirs.

REFERENCES

BELTRAMI, H., and MARESCHAL, J. C. (1993), *Strange Seismic Attractor?* Pure and Appl. Geophys. *141*, 71–81.
BHATTACHARYA, S. N., and SRIVASTAVA, H. N. (1992), *Earthquake Predictablity in Hindukush Region Using Chaos and Seismicity Pattern*, Bull Indian Soc. Earthq. Tech. *29*, 23–25.
GRASSBERGER, P., and PROCACCIA, I. (1983), *Characteristics of Strange Attractors*, Phys. Rev. Lett. *50*, 346–349.
GUPTA, H. K. *Reservoir-induced Earthquakes* (Elsevier, Amsterdam 1992).
HOROWITZ, F. G. (1989), *A Strange Attractor Underlying Parkfield Seismicity?* EOS *70*, 1359.
JULIAN, B. R. (1990), *Are Earthquakes Chaotic?* Nature *345*, 481–482.
KEBEASY, R. M., MANNMOUN, N., IBRAHIM, E., MEGAHED, A., SIMPSON, D. W., and LITH, W. S. (1987), *Earthquake Studies at Aswan Reservoir*, J. Geodyn. *7*, 173–193.
KEITH, C. M., SIMPSON, D. W., and SOBOLEVA, O. V. (1982), *Induced Seismicity and Style of Deformation at Nurek Reservoir, Tadjikistan, USSR*, J. Geophys. Res. *87*, 4609–4624.
RUELLE, D. (1990), *Deterministic Chaos: The Science and the Fiction* (The Clande Bernard Lecture), Proc. Roy. Soc. London, (Math. Phys. Sci.) *A427*, 241–248.

SIMPSON, D. W., and NEGMATULLAEV, S. K. (1981), *Induced Seismicity at Nurek Reservoir, Tadjikistan, USSR*, Bull. Seismol. Soc. Am. *71* (5), 1561–1586.

SIMPSON, D. W. (1986), *Triggered Earthquakes*, Ann. Rev. Earth Planet. Sci. *14*, 21–42.

SORNETTE, A., DUBOIS, J., CHEMINEE, J. L., and SORNETTE, D. (1991), *Are Sequences of Volcanic Eruptions Deterministically Chaotic?* J. Geophys. Res. *96*, 11931–11945.

SRIVASTAVA, H. N., RAO, D. T., and MATHURA SINGH (1991), *Seismicity Pattern for Earthquakes near Bhatsa Reservoir, India*, Tectonophys. *196*, 141–156.

SRIVASTAVA, H. N., BHATTACHARYA, S. N., and SINHA RAY, K. C. (1994), *Strange Attractor Dimension as a New Measure of Seismotectonics around Koyna Reservoir, India*, Earth and Planet. Sci Lett. *124*, 57–62.

(Received May 5, 1994, revised/accepted January 6, 1995)

PAGEOPH

Pageoph Topical Volumes

Aspects of Pacific Seismicity
Edited by
E.A. Okal
1991. 200 pages. Softcover
ISBN 3-7643-2589-5

**Experimental Techniques
in Mineral and Rock Physics**
The Schreiber Volume
Edited by
R.C. Liebermann and C.H. Sondergeld
1994. 457 pages. Softcover
ISBN 3-7643-5028-8

**Faulting, Friction, and Earthquake
Mechanics**
Edited by
C.J. Marone and M.L. Blanpied
Part I:
1994. 399 pages. Softcover
ISBN 3-7643-5073-3
Part II:
1994. 516 pages. Softcover
ISBN 3-7643-5099-7

Fractals and Chaos in the Earth Sciences
Edited by
C.G. Sammis, M. Saito and G.C.P. King
1993. 188 pages. Softcover
ISBN 3-7643-2878-9

Induced Seismicity
Edited by
A. McGarr
1993. 460 pages. Softcover
ISBN 3-7643-2918-1

**Localization of Deformation
in Rochs and Metals**
Edited by
A. Ord, B.E. Hobbs and H.-B. Mühlhaus
1991. 158 pages. Softcover
ISBN 3-7643-2772-3

**Scattering and Attenuation
of Seismic Waves**
Edited by
R.-S. Wu and K. Aki
Part I:
1988. 448 pages. Softcover
ISBN 3-7643-2254-3
Part II:
1989. 198 pages. Softcover
ISBN 3-7643-2341-8
Part III:
1990. 438 pages. Softcover
ISBN 3-7643-2342-6

**Shallow Subduction Zones: Seismicity,
Mechanics and Seismic Potential**
Edited by
R. Dmowska and G. Ekström
Part I:
1993. 220 pages. Softcover
ISBN 3-7643-2962-9
Part II:
1994. 220 pages. Softcover
ISBN 3-7643-2963-7

Tsunamis: 1992–1994
Their Generation, Dynamics, and Hazards
Edited by
K. Satake and F. Imamura
1995. 516 pages. Softcover
ISBN 3-7643-5102-0

**Please order through your
bookseller or write to:**
Birkhäuser Verlag AG
P.O. Box 133
CH-4010 Basel / Switzerland
FAX: ++41 / 61 / 271 76 66
e-mail: 100010.2310@compuserve.com

**For orders originating
in the USA or Canada:**
Birkhäuser
333 Meadowlands Parkway
Secaucus, NJ 07096-2491 / USA

Birkhäuser

**Birkhäuser Verlag AG
Basel · Boston · Berlin**